AF566834

SPIN WAVE CONFINEMENT

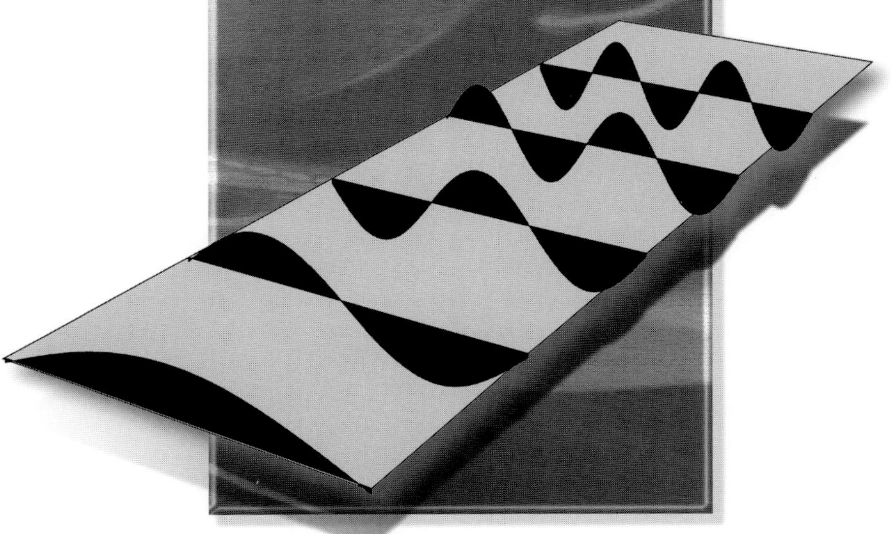

edited by

Sergej O Demokritov
University of Münster, Germany

Published by

Pan Stanford Publishing Pte. Ltd.
5 Toh Tuck Link
Singapore 596224

Distributed by

World Scientific Publishing Co. Pte. Ltd.
5 Toh Tuck Link, Singapore 596224
USA office: 27 Warren Street, Suite 401-402, Hackensack, NJ 07601
UK office: 57 Shelton Street, Covent Garden, London WC2H 9HE

British Library Cataloguing-in-Publication Data
A catalogue record for this book is available from the British Library.

SPIN WAVE CONFINEMENT

ISBN-13 978-981-4241-06-9
ISBN-10 981-4241-06-7

Printed in Singapore by B & JO Enterprise

INTRODUCTION

The concept of dynamic eigenmodes of unconfined magnetic media called spin waves was introduced by Bloch in 1930. A spin wave represents a wave of spin precession propagating in a magnetically ordered medium. Early experimental evidence for the existence of spin waves came from measurements of thermodynamic properties of ferromagnets, in particular the temperature dependence of their saturation magnetization. The first direct observation of spin precession was made using ferromagnetic resonance by Griffiths in 1946 for the case of uniform precession which can be treated as a spin wave with the zero wave vector. Later inelastic light scattering experiments performed by Fleury, Porto, Cheesman, and Guggenheim in 1966 have confirmed the existence of spin waves. In many aspects spin waves can be considered as a magnetic analogue of sound. However, as it will be discussed in this book, the main difference between spin waves and sound lies in the very different dispersion laws. Contrary to sound or light, spin waves usually have a gap in their dispersion spectrum which depends on the applied magnetic field. Thus, the propagation characteristics of spin waves can be easily manipulated by a researcher.

Being of great importance for basic research of magnetically ordered systems, spin waves play an important role for technical applications as well. In fact, since spin waves are eigenmodes of magnetic media, the dynamics of magnetization can be nicely analyzed based on the spin-wave approach. Knowing the properties of spin waves, one can obtain a vital information on the temporal response of the magnetic media to pulses of the magnetic field or microwave radiation. Apparently these properties are decisive for the functionality of magnetic sensors, memory elements, and radars.

The history of the spin-wave conception is inseparably linked with the reduction of the dimensions. Being first introduced for bulk media, spin waves in magnetic films have attracted enormous interest in 60s and 70s of the last century. It was found that the confinement caused by the finite thickness of a film results in a variety of new effects and even new types of spin waves. The so-called surface spin waves were introduced theoretically by Damon and Eshbach in 1961 and than observed experimentally by Grunberg and Metawe in 1977. Therefore, it is not surprising that the interest to spin waves in films was a forerunner of a wide application of magnetic films in the information technology.

This book on *Spin Wave Confinement* concentrates on the new physics associated with the magnetization dynamics in laterally confined magnetic structures. The topic of this book reflects the next step in reduction of the dimensionality of the magnetic medium. Instead of two-dimensional films one-dimensional and zero-dimensional objects are currently investigated very actively. Again, this trend is closely connected with the recent developments in the field of high-density magnetic storage and miniature magnetic sensors, which created much interest into the dynamic properties of one- and zero-dimensional magnetic elements. The book is aimed at the promotion of the basic knowledge in this field. On the other hand it reflects the current state of the art in this very active and fast developing research area. Three main experimental techniques used to access information on the spin-wave dynamics are presented here, such as Brillouin light scattering (Chapters 1 and 2), magneto-optical Kerr imaging (Chapters 3 and 4), and ferromagnetic resonance microwave spectroscopy (Chapters 5 and 6).

The book is structured as follows:

In the first chapter, S.O. Demokritov and V.E. Demidov present the basics equations describing spin-wave dynamics in laterally confined systems and compare the experimental results for the simple geometries (stripes and rectangular platelets) with the predictions of the theory. Surprisingly, even such a simple geometry as one-dimensional stripes demonstrates a variety of interesting effects. The final part of the chapter is devoted to nonlinear aspects of spin-wave confinement, the problem which is of great importance for practical applications.

In Chapter 2, G. Gubbiotti, M. Madami, S. Tacchi, and G. Carlotti describe spin wave confinement in more elaborated patterned structures. The lateral confinement of spin waves in long stripes patterned of magnetic multilayers is addressed. The principal effect of the inter-layer dipolar coupling in the formation of either "acoustic" (in-phase) or "optic" (anti-phase) coupled spin-wave modes is presented. More complex geometric elements, such as elliptical dots made of single magnetic films and multilayered structures are also considered in this chapter.

Chapter 3, written by V.V. Kruglyak, S. Keatley, J. Hicken, R. Childress, and J.A. Katine, reviews the spin-wave confinement effects in arrays of magnetic nanosquares with the lateral sizes down to 64 nm. It analyzes the crossover from a two mode to a single mode regime with decreasing size of the squares. Comparison of the obtained experimental data with the results of micromagnetic simulations reveals the important role of the exchange interaction for spin-wave modes in nanosquares.

In Chapter 4, M. Bailleul and Ch. Back investigate the spin-wave dynamics in magnetically unsaturated squares. One of the most often realized domain structure of magnetic squares in remanence is the well known Landau state, consisting of two crossing diagonal domain walls. The influence of multi-domain equilibrium ground state on the spin waves is investigated in detail. A less evident reverse influence of the spin-wave modes on the equilibrium structure is analyzed as well.

Chapter 5, written by D. Grundler, F. Giesen, and J. Podbielski, describes the application of the broadband GHz spectroscopy technique to the study of nanostructured permalloy rings. Nanorings are of interest for fundamental research on spin excitations in nanomagnets due to their nontrivial topology. Several intriguing spin-wave phenomena are observed: spin-wave interference in the vortex state, frequency splitting in rings with a tailored asymmetry, as well as nonlinear spin-wave effects and microwave assisted switching.

Finally, in Chapter 6, A. Slavin and V. Tiberkevich present an analytic theory of the spin-wave modes in magnetic nanocontacts excited by the spin-polarized direct current. Although the model is based on a perturbative theory of a nonlinear auto-oscillator, it gives a qualitatively correct description of the phenomenon and adequately describes most of the currently available experimental data.

The book is intended to provide a broad spectrum of information for scientists carrying out both fundamental and applied studies in the area of magnetic dynamics. It will hopefully serve as a textbook for graduate students starting their carrier in this field. I am grateful to all the authors for their contribution to the field of spin-wave confinement and wish to thank them for their cooperation during preparation of this book.

Sergej O. Demokritov
Institute for Applied Physics and
Center for Nonlinear Science
Münster University, Germany

CONTENTS

Chapter 1

QUANTIZED SPIN-WAVE MODES DUE TO LATERAL CONFINEMENT

Sergej O. Demokritov and Vladislav E. Demidov

Institute for Applied Physics and Center for Nonlinear Science, Münster University, Münster, Germany

1. Introduction

Recent developments in the field of magnetic storage and magnetic sensors created much interest to the dynamic properties of small magnetic elements.[1,2] The study of quantized and localized spin-wave modes of small magnetic elements is very important for understanding of dynamic magnetic properties of such elements. Each confined magnetic element can be considered as a magnetic resonator. The thermal noise can be observed at the resonance frequencies of such a resonator which are in the microwave frequency range. If this element, e.g., is used as a part of a magnetic reading head, or as a memory element at microwave frequencies, thermal magnetic noise limits the figures of merit of those devices.[3–5] Therefore, it is very important to know the resonance frequencies. Recently, it has been found[3–5] that the frequency spectrum of the magnetic noise is quantized, i.e. shows well defined maxima at certain frequencies. To control the noise in order to improve the properties of such laterally confined magnetic elements one needs to study the properties of thermally excited spin-wave modes in those systems.

The idea of spin-wave modes as eigen-excitations of small magnetic elements sounds intuitive and might look very simple. In fact, lateral confinement should define boundary conditions for spin-wave amplitudes at the edges of the element. These conditions provide selection rules for the wave vector of the allowed modes. Thus, instead of continuous spectrum of spin waves (the spin-wave frequency is uniquely determined by its wave vector) one obtains a series of quantized spin-wave modes, each of them is observed within a given wave-vector interval. Indications that such a straightforward scenario in fact can be observed is illustrated in Fig. 1 (from Ref. 6).

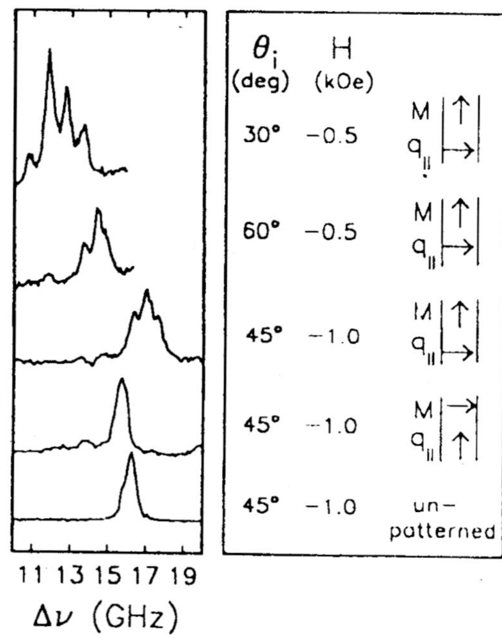

Fig. 1. Left: Spin-wave spectra of an array of NiFe stripes measured using light scattering. A spectrum of a continuous film is also given for comparison. Right: Diagrams illustrating the experimental geometry. The angle of light incidence which determines the wave vector of the studied spin wave is indicated as well.

However, the phenomenon of spin-wave confinement gains its complexity because of modification of the internal magnetic field due to patterning. In fact, since pioneering works by Kittel and Walker, who analytically derived the spin-wave spectrum of ellipsoidal elements,[7,8] and up to very recently experimentally studied patterned magnetic elements were approximated by ellipsoidal elements with corresponding demagnetizing factors.[9] Recently it has been found that a non-ellipsoidal shape of an element drastically affects its dynamic properties.[10] Because of the non-ellipsoidal shape both the static and the dynamic internal magnetic fields in the element are inhomogeneous due to demagnetizing effects. We will discuss the influence of the field inhomogeneity on the eigenmodes of the elements, gradually increasing the complexity of the problem.

First, the case of long stripes magnetized along their length will be considered. In such a stripe the static internal field is homogeneous (if one neglects the areas close to the stripe ends). The confinement shows itself in spin-wave quantization, with the component of the wave vector along the width of the stripe being quantized. Although the mean direction of the magnetization lies along the stripe axis and, therefore, does not cause a demagnetizing field, in the case of precession the magnetization tips out of the stripe plane and generates dynamic dipolar fields which oscillate with the frequency of the precession. These dipolar fields are essentially inhomogeneous, they lead to the so-called dipolar pinning.[11]

Second, for the case of a transversely magnetized stripe (i.e., a stripe magnetized along its width) the main effect is produced by a strongly inhomogeneous static internal magnetic field. This inhomogeneity causes the creation of so-called "spin-wave wells".[10,12,13] We will discuss this phenomenon in detail, analyzing the approaches for determination of the spin-wave spectrum and the distribution of the dynamic magnetization under such conditions.

Third, the more complicated case of rectangular magnetic platelets will be considered. Here confinement takes place for *both in-plane* coordinates. We will show that the combination of the effects of lateral quantization along the two coordinates as well as spatial localization of spin-wave modes takes place in this case. The problem can be theoretically analyzed using the factorization of the two-dimensional magnetization distribution in the element into one-dimensional spin-wave eigenfunctions of longitudinally and transversely magnetized stripes.

Fourth, the most intricate case of the nonlinear confinement will be considered. We will show that the nonlinear interaction between spin-wave modes makes the above factorization impossible.

The discussed experimental results are mainly obtained using Brillouin light scattering spectroscopy, which is proved to be a powerful tool for investigation of spin waves. We discuss the advantages and disadvantages of this technique in comparison with other techniques for investigation of magnetic dynamics.

This Chapter is organized as follows: In Sections 2 and 3 typical examples of magnetic patterned structures under investigation will be presented and the basics of Brillouin light scattering spectroscopy will be briefly recapitulated. These sections will be followed by Section 4, where the basic equations describing magnetic dynamics are derived. Section 5 is devoted to the simplest case of the stipes magnetized along their axes and describes the effect of spin-wave quantization in detail. Section 6 describes the spin-wave well effect and corresponding spin-wave localization in stripes, while in Section 7 the analysis of more complicated two-dimensional spin-wave confinement will be performed. The chapter will be concluded by Section 8, where influence of nonlinear effects on spin-wave confinement will be discussed followed by a summary and outlook.

2. Magnetic Confined Systems

The technology for fabrication of high-quality confined magnetic structures with lateral extensions on the micrometer, submicrometer, and nanometer length scale has been perfected to a remarkable degree in the past decades.[14–16]

Lateral magnetic structures are conveniently fabricated from magnetic films using lithographic patterning procedures. Usually, for simplicity, one considers either structures with one restricted lateral dimension (stripes), or with two restricted dimensions (circular dots or rectangular platelets) although no real reduction of dimensionality is given since in the confined elements the magnetization is not constant over each magnetic object along the directions of restricted dimension. The reason for that is the size of the samples: up to now the most of the studied samples have lateral dimensions far above the exchange correlation length, which provides a natural scale for the dimensionality crossover.

It is usually rather difficult to study a single magnetic element, since it challenges the sensitivity of the measurements setup.[17,18] To avoid this problem the elements under investigation are assembled in arrays. On the other hand, changing the distance between elements in an array, one can investigate the interaction between the elements.

Metallic Fe,[19] Ni,[20] permalloy (NiFe)[21–23] or Co[24–26] films are mainly used for patterning, although patterned elements made of yttrium iron garnet were recently used for investigation of nonlinear dynamics in confined geometries.[27] For fundamental studies permalloy films are preferable due to smallness of their coercive field and the weak intrinsic magnetocrystalline anisotropy. The saturation fields of permalloy films are small, and the vanishing intrinsic anisotropy does not inhibit the observation of sometimes minute anisotropy effects caused by, e.g., the shape of the elements or by interactions between elements. However, patterned structures made on the basis of Co films allow one to investigate the interesting case of perpendicularly magnetized dots or stripes.[24] Patterned multilayers are of particular interest since recently.[28,29] Another type of confined magnetic elements are point-contact systems.[30] Here the confinement is determined by a small area where a very dense electric current flows through a magnetic film or a multilayer.

Fig. 2. Scanning electron micrographs of the permalloy stripes with a width of 1 μm and a separation of 0.1 μm.

Figures 2 and 3 show dense arrays of magnetic stripes and rectangular platelets, correspondingly.

3. Brillouin Light Scattering Technique: A Tool for Investigation of Confined Magnetic Structures

Brillouin light scattering (BLS) spectroscopy is widely used for spin-wave studies since decades. Together with time-resolved Kerr-microscopy and x-ray microscopy BLS recently became one of the most powerful techniques for experimental investigations of inhomogeneous magnetization dynamics in confined systems. BLS has a number of advantages over other tech-

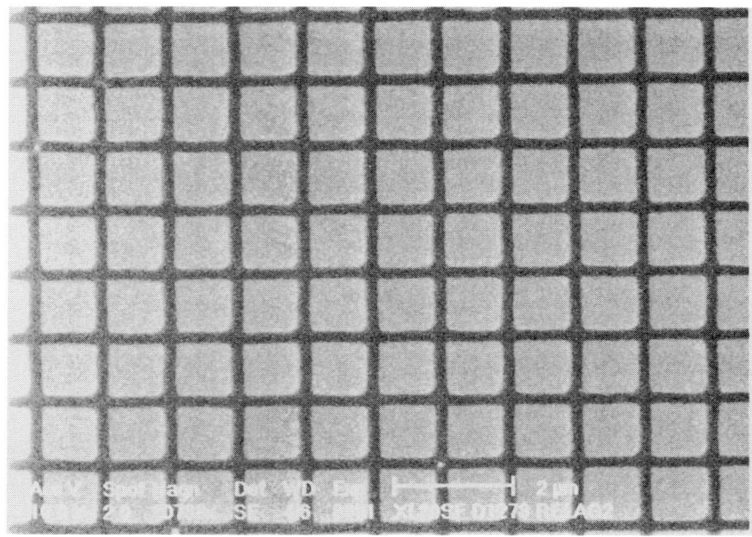

Fig. 3. Scanning electron micrograph of the array of $1 \cdot 1$ μm^2 permalloy platelets (squares) with a spacing of 0.2 μm.

niques. It provides a possibility to investigate spin waves with different absolute values and orientations of their wave vectors in a very broad frequency range from 2 GHz to 500 GHz. Moreover, an important advantage of BLS is that its sensitivity allows for detection of thermally excited spin waves, i.e. there is no need for external excitation sources. However, up to recently, the application of BLS for studies of confined magnetic structures was restricted by its poor lateral resolution defined by the size of the probing laser spot, which was usually 30–40 μm in diameter. The invention of micro-BLS (μ-BLS), where the laser beam is focused almost down to the theoretical diffraction limit of 200-250 nm and the scattered light can be effectively collected from this small area for further spectral analysis, has opened new perspectives for BLS. Magnetic dynamics of single elements with sub-micrometer sizes is accessible now.

A detailed description of BLS is published elsewhere.[31] Here we present just the basics of this technique. BLS process is illustrated in Fig. 4. Photons of energy $\hbar\omega_I$ and momentum $\hbar\vec{q}_I$ interact with the elementary quanta of spin waves $(\hbar\omega, \hbar\vec{q})$, which are magnons. The scattered photon gains an increase in energy and momentum:

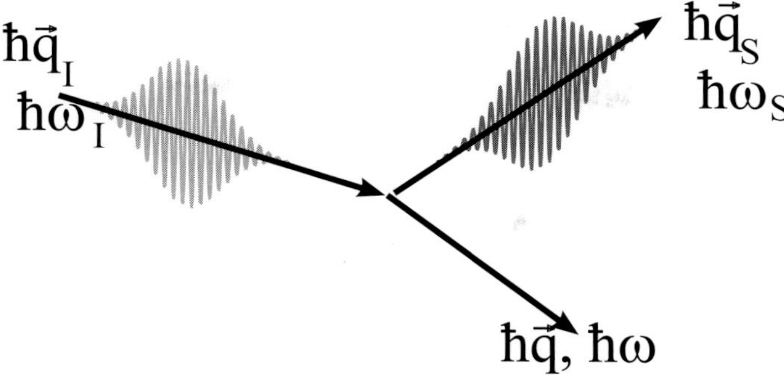

Fig. 4. Brillouin light scattering process from spin waves (magnons). The wave vectors of the incident and the scattered photons as well as that of the magnon are shown. If the magnon is absorbed, the energy of the photon increases; if the magnon is radiated, the energy of the photon is decreases.

$$\hbar \vec{q}_S = \hbar(\vec{q}_I + \vec{q})$$
$$\hbar \omega_S = \hbar(\omega_I + \omega), \qquad (1)$$

if a magnon is absorbed. From (1) it is evident, that the wave vector $\vec{q}_S - \vec{q}_I$, transferred in the scattering process, is equal to the wave vector $\vec{q}$ of the magnon. A magnon can also be created by an energy and momentum transfer from the photon, which in the scattered state has the energy $\hbar(\omega_I - \omega)$ and momentum $\hbar(\vec{q}_I - \vec{q})$. For room temperature ($T \gg \hbar\omega/k_B \approx 1K$) both processes have about the same probability. In a classical treatment the scattering process can be understood as follows: Due to magneto-optical effects a phase grating is created in the material by a spin wave. Since the grating propagates with the phase velocity of the spin wave, light is Bragg-reflected from the phase grating with its frequency Doppler-shifted by the spin-wave frequency.

The conservation law (1) for wave vectors is a consequence of the translational invariance of an infinite medium. If the translational invariance of the scattering medium is broken by confinement effects in one or more directions, the corresponding components of the wave vector, $\vec{q}$, are not fully conserved in a light scattering process. Such type of confinement should be taken into account if the incident light is strongly focused. In this case the technique looses the resolution with respect to the wave vector of the scattering wave, but wins a spatial resolution. The confinement of a spin wave itself causes uncertainty of its wave vector, which breaks the conservation law (1) as well. The uncertainty in q_i is, apparently, inversely proportional to the confinement length in the i-direction t_i. If, for example, a spin-wave mode is confined in a rectangular platelet, neither component of the wave vector is well defined. In the case of a long stripe, the component of $\vec{q}$ along the stripe axis is conserved only.

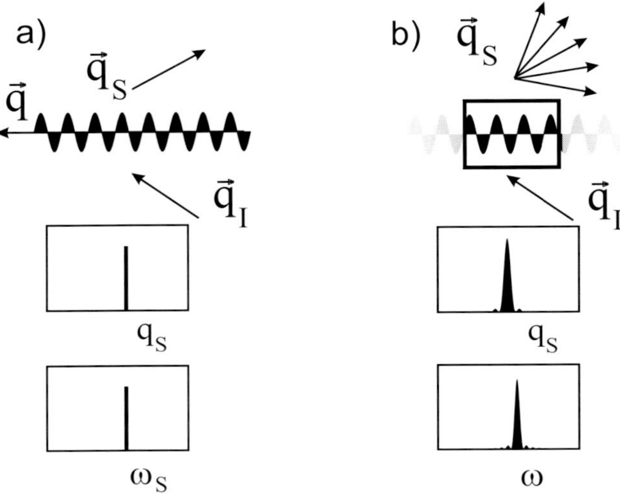

Fig. 5. Schematics of the light scattering process for (a) infinite scattering volume (plane spin wave) and (b) confined spin-wave mode.

It has been shown in,[32] that the intensity of the light scattered by spin-wave modes in a stripe as a function of the transferred wave vector q is connected with the profile of the dynamic magnetization, $m(y)$ of the spin-wave mode confined in the stripe.

$$I(q) \propto \left| \int_{-w/2}^{w/2} m(y) \times \exp(-iqy)dy \right|^2 \tag{2}$$

where w is the stripe width. Thus, the light scattering intensity is proportional to the squared Fourier transform of the dynamic magnetization, $m(y)$. As the integration is performed over the interval $[-w/2, w/2]$ and the dynamic magnetization is essentially a sine function with a wave vector k within this interval, the maximum in scattering intensity will be at $q = k$ and the broadening will be determined by the length of the interval, i.e., by the stripe width w. In the more complicated case of a rectangular element, a two dimensional Fourier transformation is needed. If BLS is used in the "Fourier microscope" mode, the light scattering intensity is measured as a function of $\mathbf{q}$. On the basis of this information the mode profile $m(\mathbf{r})$ in the elements can be reconstructed. In the "Fourier microscope" mode the spatial resolution, δ is determined by the accessible transferred wave vector interval $\pi/\Delta q$. For backscattering experiments it follows that $\Delta q = 2q_I$ and $\delta = 120$–130 nm for green visible light.

Note here, that similar arguments can be used for the analysis of the energy conservation, which is a consequence of the time invariance of the problem. If, however, either the incident laser beam or the spin wave have a finite duration T, the uncertainty of the frequency/energy is proportional to $1/T$.

In the BLS studies light of a single-frequency laser is focused onto the sample by an objective lens. The light scattered from the sample (elastic and inelastic contributions) is collected and sent to a Fabry-Perot interferometer.[33,34] In order to obtain the high contrast necessary to detect the weak inelastic signals, the light passes the interferometer several times guided by a system of retroreflectors and mirrors. The frequency selected light transmitted by the interferometer is detected by a photomultiplier. Data collection is performed by a multichannel analyzer realized using a personal computer.

A recent development of the BLS-technique is the so-called micro-BLS, allowing direct spatial resolution close to 250 nm.[18] To obtain a small focal spot of the incident beam, it is of great importance to apply a laser beam corresponding to a single spatial mode and to reduce the divergence of the beam. For this purposes an optical system including a beam expander and a spatial filter is usually used. The resulting beam of the TEM_{00} spatial mode is focused by a microscope objective onto the surface of the sample placed into a magnetic field. The objective allows focusing of TEM_{00} beams with small divergences down to about 250 nm and effective collection of the light scattered from this tiny area for further spectral analysis. The sample is mounted on a xyz piezoelectric stage, which provides the sample positioning along all three dimensions with a precision of about 20 nm. To detect the inelastically scattered light the spectrum of the scattered light is analyzed by means of the standard BLS-technique based on a Fabry-Perot interferometer. Taking into account the value of the numerical aperture of the objective and the uncertainty in the wave vector of the scattered light caused by the spatial confinement, the micro-BLS setup does not provide any wave vector selectivity. Instead the contributions from all spin waves up to a certain wave vector are automatically integrated. One can easily estimate the cut-off wave vector, limiting the sensitivity of the setup: $q_{max} = (1.5 - 2.5) \cdot 10^5$ cm^{-1}, depending on the wavelength of the laser beam. Thus, one can simultaneously collect a BLS-signal from all spin waves propagating in different directions and having their wave vectors $q < q_{max}$. For investigation of magnetic nanostructures it is extremely important to place the laser spot on the appropriate element of the structure and to keep it at the same place over the measurement time. For a visual control of the position of the laser spot on the surface of the sample, a viewing system is integrated into the setup. This system consists of a CCD camera supplied with a telescopic objective, two beam splitters, and a source of white light. The system allows direct observation of the surface of the sample and the position of the probing laser spot on the screen of a monitor during measurements. In extreme cases an active beam stabilization technique, based on a digital analysis of the obtained image and linked with the piezoelectric stage, can be applied.

4. Theory of Spin-Wave Modes in Confined Geometries

In this section we give a brief introduction to the theory of confined spin-wave modes. Since the lateral confinement introduces an essential inhomogeneity of the internal magnetic field even in a homogeneously magnetized sample, the resulting effect of the inhomogeneous static internal field and that of the inhomogeneous dynamic dipole fields is of great importance.

To limit the complexity of the problem we will mainly consider quasi-one-dimensional objects (long stripes), magnetized either longitudinally or transversally. The two discussed geometries and the corresponding coordinate systems are schematically shown in Fig. 6.

The magnetization dynamics of a ferromagnet is described by the Landau-Lifshitz torque equation[35]

$$-\frac{1}{\gamma}\frac{d\mathbf{M}}{dt} = \mathbf{M} \times \mathbf{H}_{\text{eff}} \,, \tag{3}$$

where $\mathbf{M}$ is the total magnetization, γ is the modulus of the gyromagnetic ratio for the electron spin ($\gamma/2\pi = 2.8\,\text{MHz/Oe}$), and $\mathbf{H}_{\text{eff}} = -\delta W/\delta\mathbf{M}$ is the effective magnetic field calculated as a variational derivative of the energy function W, where all the relevant interactions in the magnetic substance have been taken into account (see, e.g. Ref. 36).

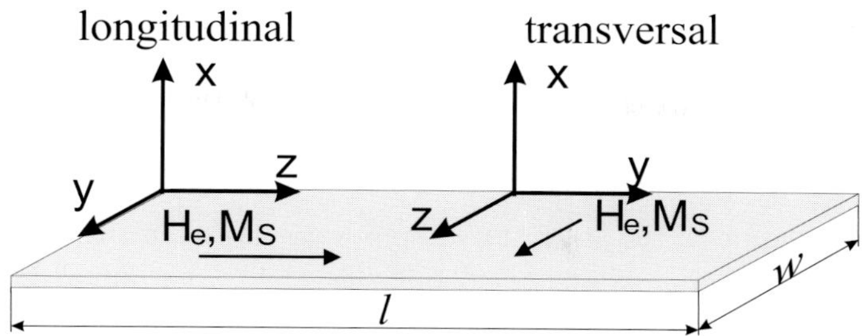

Fig. 6. Used coordinate system for the longitudinal and the transverse field geometries. Note that the z-axis is parallel to the direction of the external field $\mathbf{H}_e$ and the static magnetization $\mathbf{M}_S$. The lateral dimensions of the stripe $w \ll l$ are indicated as well.

We assume that due to applied external magnetic field $\mathbf{H}_e = H_e\mathbf{e}_z$ the static magnetization in the elements is uniform $\mathbf{M}_{stat} = M_S\mathbf{e}_z = \text{const}$ everywhere, except the narrow regions close to the edges perpendicular to the applied field (edge domains). Both crystalline and surface anisotropy of the material are neglected, which is a good approximation for permalloy. We also consider only spin-wave modes having a uniform distribution of the dynamic magnetization $\mathbf{m}$ along the x-axis. $\mathbf{m}(\mathbf{r}, t) \to \mathbf{m}(\vec{\rho}, t)$, where $\vec{\rho} = y\mathbf{e}_y + z\mathbf{e}_z$. The so-called perpendicular standing spin waves (PSSW) with a non-uniform distribution of $\mathbf{m}$ are much easier for the analysis, since they are not affected by the lateral confinement. The frequencies of PSSW are determined by the isotropic exchange interaction and the external magnetic field.[31]

Theory of nonlinear waves is rather complex, extended analysis of this topic can be found in Ref. 37. A specific section of this Chapter is devoted to nonlinear spin-wave effects for confined geometries. However, in this Section we will neglect the nonlinear effects, i.e., we restrict ourselves by the case $m \ll M_S$. Taking into account the above simplifications, the magnetization vector and the effective field can be expressed as

$$\mathbf{M}(\vec{\rho}, t) \approx M_S\mathbf{e}_z + m_x(\vec{\rho}, t)\mathbf{e}_x + m_y(\vec{\rho}, t)\mathbf{e}_y \,, \tag{4}$$

$$\mathbf{H}_{\text{eff}}(\vec{\rho}, t) = \mathbf{H}_e + \frac{2A}{M_S^2}\nabla_\rho^2\mathbf{M}(\vec{\rho}, t) + 4\pi \int \hat{\mathbf{G}}(\vec{\rho}, \vec{\rho}') \cdot \mathbf{M}(\vec{\rho}', t) d\vec{\rho}' \,, \tag{5}$$

where A is the exchange constant and $\nabla_{\vec{\rho}}^2 = \frac{\partial^2}{\partial y^2} + \frac{\partial^2}{\partial z^2}$ is the two dimensional Laplace operator. $\hat{\mathbf{G}}(\vec{\rho}, \vec{\rho}')$ is the the tensorial Green's function of the dipolar interaction, as discussed below. The integral in (5) describes the static and dynamic demagnetizing fields. The effective field $\mathbf{H}_{\text{eff}}(\vec{\rho}, t)$

can be also presented as a sum of a static and a dynamic part similar to the presentation of the magnetization (4),

$$\mathbf{H}_{\text{eff}}(\vec{\rho},t) = H_{\text{i}}(\vec{\rho})\mathbf{e}_z + h_{\text{eff},x}(\vec{\rho},t)\mathbf{e}_x + h_{\text{eff},y}(\vec{\rho},t)\mathbf{e}_y \ . \tag{6}$$

with $h_{\text{eff}} \ll H_{\text{i}}$.

For calculation of the spectrum of spin-wave modes in the element the formalism of tensorial Green's functions developed for infinite in-plane magnetized films in[38] will be used. In the framework of this formalism the linearized Landau-Lifshitz equation and the Maxwell equations in the magnetostatic limit are reduced to an integro-differential equation for the dynamic magnetization $m_x(\vec{\rho},t) = m_x(\vec{\rho}) \times \exp(i\omega t)$ and $m_y(\vec{\rho},t) = m_y(\vec{\rho}) \times \exp(i\omega t)$. Inserting (4) and (5) into the Landau-Lifshitz equation (3) and taking into account only the linear terms in m_x and m_y one obtains the commonly used form

$$\begin{aligned} &\left[-\alpha\omega_M \nabla^2_{\vec{\rho}} + \omega_{H_{\text{i}}}(\vec{\rho})\right] \hat{\mathbf{I}}\mathbf{m}(\vec{\rho}) + i\omega\hat{\mathbf{T}}\mathbf{m}(\vec{\rho}) \\ &\quad -\omega_M \textstyle\int d\vec{\rho}\,'\hat{\mathbf{G}}_{x,y}(\vec{\rho},\vec{\rho}\,')\mathbf{m}(\vec{\rho}\,') = 0 \ . \end{aligned} \tag{7}$$

Here

$$\hat{\mathbf{I}} = \begin{bmatrix} 1 & 0 \\ 0 & 1 \end{bmatrix}, \hat{\mathbf{T}} = \begin{bmatrix} 0 & -1 \\ 1 & 0 \end{bmatrix}, \hat{\mathbf{G}}_{x,y} = \begin{bmatrix} G_{xx} & G_{xy} \\ G_{yx} & G_{yy} \end{bmatrix}, \mathbf{m} = \begin{pmatrix} m_x \\ m_y \end{pmatrix} \tag{8}$$

and $\omega_M = \gamma 4\pi M_{\text{S}}$, $\alpha = \frac{A}{2\pi M_{\text{S}}^2}$ is the exchange stiffness constant,

$$\omega_{H_{\text{i}}}(\vec{\rho}) = \gamma H_{\text{i}}(\vec{\rho}) \ , \tag{9}$$

with the internal field

$$H_{\text{i}}(\vec{\rho}) = H_{\text{e}} - N_{zz}(\vec{\rho}) \times 4\pi M_{\text{S}} \ , \tag{10}$$

and the coordinate-dependent demagnetizing factor for a rectangular prism[9]

$$N_{zz}(\vec{\rho}) = -\int d\vec{\rho}\,' G_{zz}(\vec{\rho},\vec{\rho}\,') \tag{11}$$

Equation (7) is a two-dimensional integro-differential eigenequation for the dynamic magnetization $\mathbf{m}(\vec{\rho})$. For the analysis of (7) the components of the tensorial dipolar Green's function for a platelet should be derived from the most general form $\hat{\mathbf{G}}(\mathbf{r},\mathbf{r}') = [G_{\alpha\beta}]$, $\alpha,\beta \in \{x,y,z\}$ with

$$G_{\alpha\beta}(\mathbf{r},\mathbf{r}') = -\frac{1}{4\pi}\frac{\partial}{\partial\alpha}\frac{\partial}{\partial\beta'}\frac{1}{|\mathbf{r}-\mathbf{r}'|} \ . \tag{12}$$

Since the magnetization is assumed to have a uniform distribution across the thickness of the sample L, (12) can be simplified by averaging across the thickness:

$$G_{\alpha\beta}(\vec{\rho}, \vec{\rho}') = \frac{1}{L} \int_{-L/2}^{L/2} \int_{-L/2}^{L/2} G_{\alpha\beta}(\mathbf{r}, \mathbf{r}') dx dx' \ . \tag{13}$$

General expressions for the components of the Green's function are presented in.[39] Here we consider several special case important for the analysis of experimental results.

4.1. *Longitudinally magnetized long stripes*

Let us consider a stripe of the length l which is much larger then the width of the stripe w, magnetized along its length, which defines the z direction. The spin-wave modes have no confinement along this direction. For simplicity consider mode with the wave vector component $k_z = 0$, defining the wave vector of the wave perpendicular to the applied magnetic field, as typical for the Damon-Eshbach geometry.[40]

In that case the problem becomes one-dimensional, i.e. translationally invariant along the z direction, reducing the Green's function to $G_{\alpha\beta}(y, y')$:

$$\hat{\mathbf{G}}(y, y') = \begin{pmatrix} G_{xx} & 0 & 0 \\ 0 & G_{yy} & 0 \\ 0 & 0 & 0 \end{pmatrix} \tag{14}$$

with

$$G_{xx}(y, y') = \frac{1}{2\pi L} \ln \left[\frac{(y - y')^2}{(y - y')^2 + L^2} \right] , \tag{15}$$

$$G_{yy}(y, y') = -\delta(y - y') - G_{xx}(y, y') \ . \tag{16}$$

Since $G_{zz}(y, y') = 0$, $N_{zz}(\vec{\rho}) = N_{zz}(y) = 0$ and the internal static field $\mathbf{H}_{\mathrm{i}} = H_{\mathrm{i}}\mathbf{e}_z$ coincides with the external field $\mathbf{H}_{\mathrm{e}} = H_{\mathrm{e}}\mathbf{e}_z$. The equation of motion (7) is accordingly reduced to

$$\left[-\alpha\omega_M \frac{d^2}{dy^2} + \omega_H \right] \hat{\mathbf{I}}\mathbf{m}(y) + i\omega\hat{\mathbf{T}}\mathbf{m}(y) - \omega_M \int_{-\frac{w}{2}}^{\frac{w}{2}} dy' \hat{\mathbf{G}}_{x,y}(y, y')\mathbf{m}(y') = 0 \ . \tag{17}$$

Let us emphasize here, that despite the static demagnetizing field is vanishing ($G_{zz}(y,y')=0$) and the static internal field is homogeneous, the dynamic demagnetizing field described by the integral term in (17) is substantially inhomogeneous. At the stripe edges ($y=\pm\frac{w}{2}$), the dynamic demagnetizing field strongly grows resulting to additional pinning of the magnetization **m** at the edges.[11]

Since a typical width of the stripes, used in the experiment ($w\cong 1\mu$m) is much larger than the exchange length in 3d-metals ($l_{ex}=\sqrt{\alpha}\cong 5-10$ nm), the exchange term in (17) can be neglected and the eigenfunctions $\varphi_m(y)$ of the dipolar integral operator (last term in (17)) determined as

$$\int_{-\frac{w}{2}}^{\frac{w}{2}} dy' 4\pi \mathbf{G}_{xx}(y,y')\varphi_m(y') = \lambda_m \varphi_m(y)\,. \tag{18}$$

can be used as probe functions for solution of (17) giving

$$\left(\frac{\omega_m}{\omega_M}\right)^2 = \left(\frac{\omega_H}{\omega_M}+\alpha\kappa_m^2+1+\frac{\lambda_m}{4\pi}\right)\left(\frac{\omega_H}{\omega_M}+\alpha\kappa_m^2-\frac{\lambda_m}{4\pi}\right)\,. \tag{19}$$

To obtain the dispersion equation for the spin-wave modes of the stripe, the eigenvalues λ_m should be determined using the boundary conditions for the variable magnetization at the lateral edges of the stripe:[11]

$$\pm\frac{\partial\varphi_m(y)}{\partial y}+d(p)\varphi_m(y)\bigg|_{y=\pm\frac{w}{2}}=0\,, \tag{20}$$

where the effective "pinning" parameter:

$$d(p)=\frac{2\pi}{p[1+2\ln(1/p)]} \tag{21}$$

depends on the stripe aspect ratio $p=\frac{L}{w}\ll 1$.

It was also shown in Ref. 11 that for $p\ll 1$ the symmetric eigenfunctions $\varphi_m^s(y)$ and the antisymmetric eigenfunctions $\varphi_m^a(y)$ can be approximated by the cosine- and sine-functions

$$\varphi_m^s(y)=A_m^s\cos(\kappa_m y),\qquad \varphi_m^a(y)=A_m^a\sin(\kappa_m y)\,, \tag{22}$$

where

$$\kappa_m = m\frac{\pi}{w}\left(1-\frac{2}{d(p)}\right),\quad m=1,2,3,\ldots \tag{23}$$

Correspondingly

$$\lambda_m(p) = -4\pi + \pi^2(2m-1)p, \quad m = 1, 2, 3, \ldots \tag{24}$$

Substituting (23) and (24) in (19) one obtains the expression for the discrete frequencies of the spin-wave modes in a longitudinally magnetized stripe, which we do not present here due to its complexity. If, however, the exchange contribution and small terms proportional to p^2 ($p = \frac{L}{w} \ll 1$) are ignored, the solution can be simplified as:

$$\omega_m^2 = \omega_H(\omega_H + \omega_M) + \omega_M^2 \frac{\pi(2m-1)}{4} p, \quad m = 1, 2, 3, \ldots \tag{25}$$

4.2. *Transversely magnetized long stripes*

In this subsection we consider a more complex case of a stripe magnetized along its width w. The complexity of the problem is connected with the fact that in this case the static demagnetizing field is not zero and is inhomogeneous. To restrict the complexity of the problem let us consider only the solutions which are translationally invariant along the length of the stripe. In this case an one-dimensional analysis can be applied as well.

In the transverse geometry (7) can be rewritten as:

$$\left[-\alpha\omega_M \frac{d^2}{dz^2} + \omega_{H_\mathrm{i}}\right] \hat{\mathbf{I}}\mathbf{m}(z) + i\omega\hat{\mathbf{T}}\mathbf{m}(z) - \omega_M \int_{-\frac{w}{2}}^{\frac{w}{2}} dz' \hat{\mathbf{G}}_{x,y}(z, z')\mathbf{m}(z') = 0\,. \tag{26}$$

In contrast to (17) the quantity $\omega_{H_\mathrm{i}}(z) = \gamma H_\mathrm{i}(z)$, proportional to the internal magnetic field $H_\mathrm{i}(z)$, is substantially dependent on the coordinate z along the stripe width w due to the inhomogeneous static demagnetizing field. This inhomogeneous field can be obtained assuming homogeneous magnetization[9]

$$H_\mathrm{i}(z) = H_\mathrm{e} - N_{zz}(z) \times 4\pi M_\mathrm{S}\,, \tag{27}$$

with

$$N_{zz}(z) = \frac{1}{\pi} \times \left[\arctan\left(\frac{L}{2z+w}\right) - \arctan\left(\frac{L}{2z-w}\right)\right]. \tag{28}$$

Note here that the above expression is not valid close to the edges of the stripe, since in this regions the magnetization is strongly inhomogeneous: it rotates toward the direction parallel to the edge.

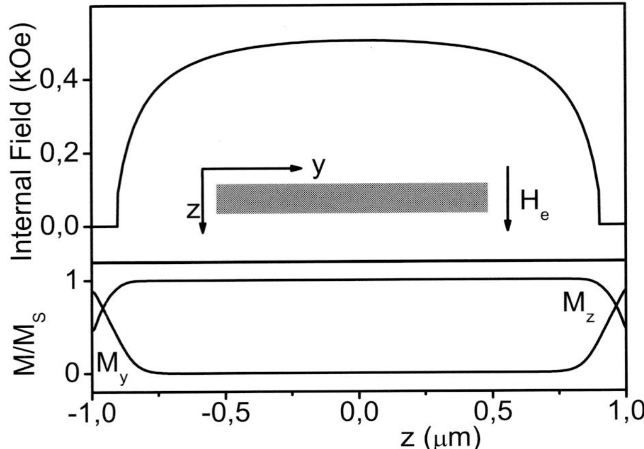

Fig. 7. Upper panel: Internal magnetic field for a transversely magnetized permalloy stripe with a width 2 μm and a thickness of 30 nm in an applied field of 0.6 kOe. The inset illustrates the geometry. Lower panel: The corresponding distribution of the static magnetization.

The general Green's function $\hat{\mathbf{G}}(\vec{\rho}, \vec{\rho}')$ for this particular case is given by

$$\hat{\mathbf{G}}(z, z') = \begin{pmatrix} G_{xx} & 0 & 0 \\ 0 & 0 & 0 \\ 0 & 0 & G_{zz} \end{pmatrix} , \tag{29}$$

with

$$G_{xx}(z, z') = \frac{1}{2\pi L} \ln \left[\frac{(z - z')^2}{(z - z')^2 + L^2} \right] , \tag{30}$$

$$G_{zz}(z, z') = -\delta(z - z') - G_{xx}(z, z') . \tag{31}$$

The inhomogeneity of the internal field H_{i} (see (27) and (28)) plays an important role for the problem, therefore let us discuss it in more detail. The field profile $H_{\mathrm{i}}(z)$ calculated using (27) and (28) is illustrated in Fig. 7 together with a sketch illustrating the reorientation of the static magnetization within the stripe. It is seen from the figure that in the mid region of the stripe where the magnetization is parallel to the applied field the internal field is almost uniform. However, close to the edges of the stripe where the magnetization direction rotates, the internal field decreases essentially. In the dipole approximation the internal field in the edges domains is zero.[41] It can be understood based on the symmetry considerations: there is no static torque, exerted on the magnetization by the internal field, i.e., the cross product $\mathbf{M} \times \mathbf{H}_\mathrm{i} = 0$, otherwise the magnetization would rotate. The dipole internal field $\mathbf{H}_\mathrm{i}$ is oriented along z as the distribution of div$\mathbf{M}$ possesses a translational symmetry along y. Since the magnetization within the edge domain is not parallel to z, the only possibility for the torque to be zero is $\mathbf{H}_\mathrm{i} = 0$. Taking into account the exchange contribution to the effective internal field results in a non-zero total effective internal field in the edges domain, but in most of the cases the effect is small.

Below we will analyze the spin-wave dynamics between the edge domains only. This region is isolated from the edge domains, since spin waves approaching the domains from the mid of the stripe are reflected by the region of a strong changes of both the internal field and the magnetization. Between the edge domains the expression for the internal field (27) with a coordinate-dependent demagnetizing coefficient (28) can be used. Using the obtained profile of the internal field and assuming a homogeneous distribution of the static magnetization, spin-wave modes in the region can be obtained as a solution of (26) using the dipole boundary conditions (20) for the dynamic magnetization. Since the stripe thickness L is much smaller than the stripe width w ($L \ll w$), the boundary conditions for the lowest spin-wave modes are very close to the boundary conditions of the total pinning for the variable magnetization at the edges $m|_{z=\pm\frac{w}{2}} = 0$.

The eigenfunctions $\mu_n(z)$ of the variable magnetization m_y describing the obtained spin-wave modes are presented in Fig. 8 for $H_\mathrm{e} = 2000$ Oe. As seen from the figure there are two different types of spin-wave modes in the spectrum: localized modes existing in the narrow regions near the stripe edges (numbered with the index n), and delocalized quasi-cosinusoidal modes covering the entire region (numbered with the index n'). The localization of the modes happens due to the spin-wave well (SWW)-effect.[10,12] The well is built close to the edge domain. The region of the strong inhomogeneity of the internal field forms one side of the well, while the edge domain forms the other side. Since the localization regions close to the edges of the stripe are nicely separated from each other, the localized modes appear in Fig. 8 as a nearly frequency degenerate doublet consisting of a symmetric and an anti-symmetric mode. The complete degeneracy is broken by the weak dynamic dipolar interaction between the parts of the modes localized close to the left and the right edges of the stripe.

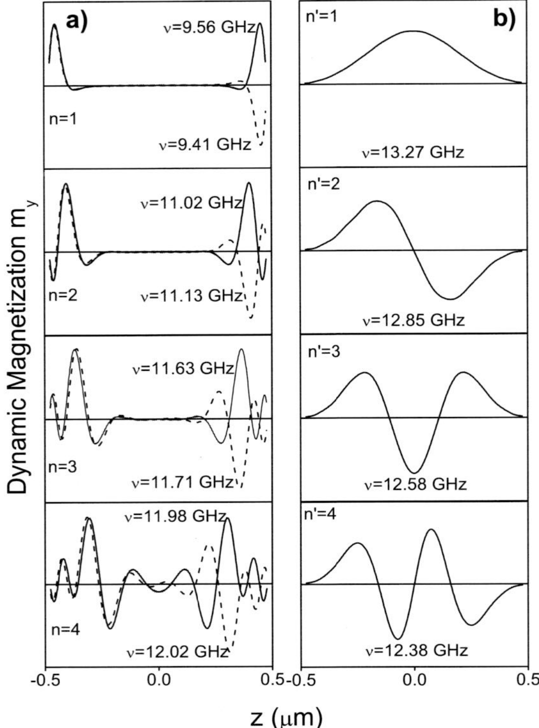

Fig. 8. Mode profiles and corresponding frequencies of an infinitely long, 1 μm wide, and 33 nm thick stripe in the transverse geometry. $H_{\rm e}$=2000 Oe. The mode profiles $m_y(z)$ were obtained by solving (26) numerically. The index n denotes the exchange dominated modes, while n' denotes the dipolar-dominated modes as discussed in the text. Note, that since the exchange-dominated modes are localized in the two regions close to the edge domains, they build nearly frequency-degenerated doublets.

To understand the SWW-effect[10,12] let us consider the dispersion law for dipole-exchange spin waves, having a wave vector parallel to the magnetic field (the so-called magnetostatic backward volume waves (MSBVW)) in an infinite film,[38] which is given by

$$\left(\frac{\omega}{\omega_M}\right)^2 = \left(\frac{\omega_H}{\omega_M} + \alpha q^2\right)\left(\frac{\omega_H}{\omega_M} + \alpha q^2 + 1 - P(qL)\right) . \tag{32}$$

where $\omega_H = \gamma H_{\rm i}$ and

$$P(qL) = 1 - \frac{1 - \exp{(qL)}}{qL} . \tag{33}$$

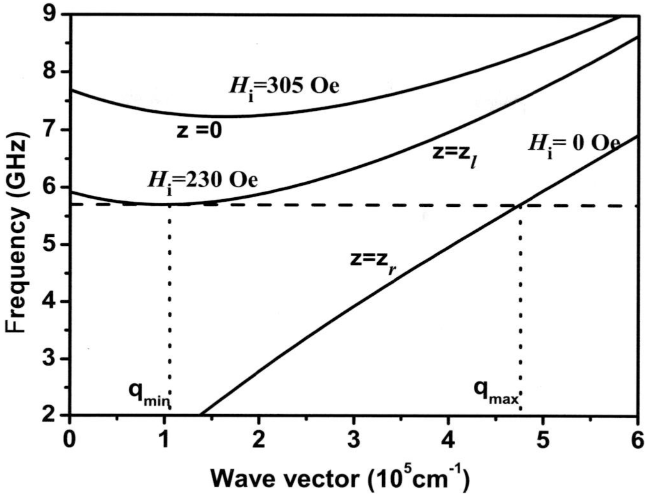

Fig. 9. Dispersion of spin waves in the MSBVW-geometry at different values of the internal field as indicated, calculated using (32). For the meaning of the wave vectors q_{min} and q_{max} see the main text.

is the dipole matrix element. The above dispersion is shown in Fig. 9 for several values of the internal field H_i

As discussed above, the internal field $H_\mathrm{i}(z)$ in a transversely magnetized stripe is strongly inhomogeneous: it is zero close to the edge domains and reaches its maximum value $H_\mathrm{i}^{max} = H_\mathrm{e} - N_{zz}(z=0)\cdot 4\pi M_\mathrm{S}$ in the middle of the stripe. Let us consider a spin wave, propagating from a point z_r in the edge domain towards the center of the stripe (wave vector $\mathbf{q}||\mathbf{M}_\mathrm{S}||\mathbf{e}_z$) in the region the inhomogeneous internal static magnetic field $H_\mathrm{i} = H_\mathrm{i}(z)$. The wave keeps propagating through the inhomogeneous field, albeit with changing wave vector, $q = q(z)$ to fulfill the dispersion $\omega = \omega(q(z), H_\mathrm{i}(z))$.[42] However, it will be reflected at the point $z = z_l$, where H_i becomes so large ($H_\mathrm{i} = 230$ Oe in our example), that no solution $\omega = \omega(q, H_\mathrm{i})$ with a real q exists anymore. The value q_{min} indicates the minimum value of the wave vectors where such an accommodation can be performed. The spin-wave mode is localized in the spin-wave well region of a rapidly changing internal field near the stripe edge. The profiles of these localized modes are illustrated in panel (a) of Fig. 8. If the frequency of the spin wave is high enough that the critical field is above the maximum internal field, the wave is not localized and can fill almost the entire region between the edge domains, as it is shown in panel (b) of of Fig. 8.

Similar to the case of the longitudinal geometry the frequencies of the spin-wave modes in transversally magnetized stipes are quantized. However, the quantization conditions for the latter case are more complicated than (23), again due to the inhomogeneous internal field. The analogue of (23) can be written in the integral form:

$$\Delta\phi_l + \Delta\phi_r + 2 \times \int_{z_l}^{z_r} q(H_{\mathrm{i}}(z), \omega) dz = 2n\pi, \tag{34}$$

where $n = 1, 2, 3, \ldots$ is the localized mode index. $\Delta\phi_l, \Delta\phi_r \in [0, \pi]$ are the phase jumps at the left and right turning points z_l and z_r of the localization interval, between which (32) has a real solution $q(z)$ for a fixed frequency ω_1. Henceforth, we will call the integral in (34) a quantization integral. The physical meaning of the quantization integral is rather obvious: the total phase, acquired by the wave on its way from z_l to z_r and back should be an integer of 2π.

The calculation of the phase jumps at the turning points is not a trivial problem. In fact, equation (34) is formulated using the WKB approximation, which is widely applied in quantum mechanics for the quasi-classical adiabatical description. In this method it is assumed that $m(z)$ is proportional to $\exp(ik(z)z)$ with a position-dependent wave vector $k(z)$, which is determined from (32) for the constant frequency of the wave mode. To calculate the phase jumps one needs to consider the reflection of the wave at the turning points beyond the WKB approximation.

Finally, we would like to stress, that the above analysis was made under the assumption that the direction of the static magnetization is parallel to the direction of the applied field everywhere in the stripe. This assumption is a good approximation for strong external fields. In a small applied field, comparable to the demagnetizing field in the center of the stripe, the magnetization rotates across the whole width of the stripe. The spin-wave modes in a transversely magnetized stripe in the case of the inhomogeneous magnetization were discussed in Refs. 44–47.

4.3. *Rectangular platelets*

The eigenfrequencies of spin-wave modes and their two-dimensional eigenfunctions can be obtained as numerical solutions of (7). However, technically, this approach is not much simpler than a direct micromagnetic numerical simulation using, for example, Langevin dynamics.[10] A different approach is to create an approximative solution describing such modes using the insight obtained from the analysis of the two above described one-dimensional eigenvalue problems for longitudinally and transversely magnetized long stripes. This approach is based on the assumption, that the two-dimensional eigenfunction, i.e., the distribution of the dynamic magnetization $m_{m,n}(y,z)$ in a spin-wave eigenmode of the platelet can be factorized as

$$m_{mn}(y,z) = \varphi_m(\kappa_{my}y) \times \mu_n(\kappa_{nz}z), \tag{35}$$

where φ_m are the cos-like eigenfuntions of a longitudinally magnetized stripe, while the functions μ_n are the eigenfunctions of a transversely magnetized stripe shown in Fig. 8. Here κ_{my} and κ_{nz} are the quantized values of the corresponding projections of spin-wave wave vectors, and $m,n = 1,2,3,\ldots$. This assumption is not justified close to the corners of the platelets. However, the regions where the assumption is not valid shrink with decreasing thickness of the platelet. Therefore, it is a good approximation for thin platelets.

The quantized values of the wave vector projection κ_{my} perpendicular to the applied field can be easily determined from (23). The determination of κ_{nz} is not so simple, since the quantization condition in a transversely magnetized stripe are defined by the integral (34) and the wave vector varies with z. However, one can introduce the mean value of the wave vector of each mode, which can be evaluated for a localized mode as

$$\kappa_{nz} = \frac{n\pi}{\Delta z_n}\,, \tag{36}$$

where $\Delta z_n = z_l - z_r$ is the localization length for a particular mode. As it has been shown both theoretically and experimentally,[10,12,13] the localization length of the modes localized close to the edges can be less than $\frac{1}{10}$ of a stripe width $\Delta z < \frac{w}{10}$, so that the characteristic wave vector of the mode along the direction of magnetization can be rather high. For example, for $w = 1\mu$m it exceeds the maximum value of the wave vector accessible by the BLS experimental technique (usually 2.5×10^5 cm^{-1}). An alternative approximate analytic treatment of the localized spin-wave modes which approximates the internal field profile by a cosine and localized modes by Mathieu functions is given in Ref. 48.

Using eigenfunctions (35) one can represent the variable magnetization of a platelet in the form of a series of eigenfunctions

$$\mathbf{m}(y,z) = \sum_{m,n} \mathbf{A}_{mn}\varphi_m(\kappa_{my}y) \times \mu_n(\kappa_{nz}z)\,. \tag{37}$$

Substituting (37) for $\mathbf{m}(y, z)$ in (7) and evaluating matrix elements of the Green's function $\hat{\mathbf{G}}_{yz}$, we can obtain an approximative dispersion equation for the discrete spin-wave modes of the platelet.

There is, however, a way to estimate these frequencies using the fact that all the matrix elements of the dynamic dipole-dipole interaction in a thin magnetic element ($L/l \ll 1$, $L/w \ll 1$) contain small aspect ratios of the element L/l or L/w. This allows to use the dispersion equation for an infinite magnetic film where the continuous wave vector of the plane wave is substituted by the quantized values κ_{my} and κ_{nz}.

An important qualitative feature that distinguishes a platelet from an infinite film is the inhomogeneity of the internal field inside the element. This inhomogeneity leads to different effective values of the internal field for different spin-wave modes, which can be calculated using the mode distributions (35) and the coordinate dependent effective demagnetizing factors (10).

Finally, taking into account the above arguments the dispersion equation can be written as:

$$\omega_{mn}^2 = \left(\omega_H^{mn} + \alpha\omega_M\kappa_{mn}^2\right)\left(\omega_H^{mn} + \alpha\omega_M\kappa_{mn}^2 + \omega_M F_{mn}(\kappa_{mn}L)\right) . \quad (38)$$

Here ω_H^{mn} is defined as

$$\omega_H^{mn} = \omega_H - \omega_M N_{mn} , \quad (39)$$

where

$$N_{mn} = \frac{4}{wl}\int d\vec{\rho}\, m_{mn}^2(\vec{\rho})N_{zz}(\vec{\rho}) . \quad (40)$$

The matrix element of the dipole-dipole interaction $F_{mn}(\kappa_{mn}L)$ is written as follows

$$F_{mn}(\kappa_{mn}L) = 1 + P(\kappa_{mn}L)\left[1 - P(\kappa_{mn}L)\right]\left(\frac{\omega_M}{\omega_H^{mn} + \alpha\omega_M\kappa_{mn}^2}\right)\left(\frac{\kappa_{my}^2}{\kappa_{mn}^2}\right) - P(\kappa_{mn}L)\left(\frac{\kappa_{nz}^2}{\kappa_{mn}^2}\right) \quad (41)$$

with $P(\kappa_{mn}L)$ defined by (33) and $\kappa_{mn}^2 = \kappa_{my}^2 + \kappa_{nz}^2$.

5. Spin-Wave Quantization in Longitudinally Magnetized Stripes

The main effect which follows from the spatial confinement in this geometry is the quantization of spin waves caused by the physical boundaries of

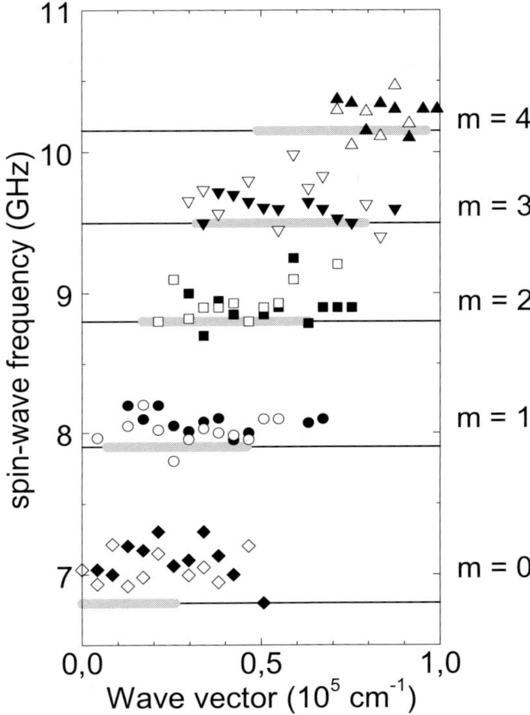

Fig. 10. Frequencies of spin-wave modes obtained from BLS spectra for longitudinally magnetized stripes of 1.8 μm width in an external field of 500 Oe. Stripe separations are 0.7 μm (full symbols) and 2.2 μm (open symbols). The horizontal lines correspond to the calculated frequencies of the five lowest-order quantized spin-wave modes. The gray bars indicate the calculated wave vector regions of observability of the discrete modes.

the stripe. In fact, if the stripe is magnetized longitudinally, the internal magnetic field is homogeneous and equal to the applied external field, if the ends of the stripes are neglected. The effect of lateral confinement due to a final width of the stripe is mostly important for the spin waves with their wave vector perpendicular to the field. Since the Damon-Eshbach (DE) modes in an unconfined film correspond to this experimental geometry, it is sometimes called as the Damon-Eshbach geometry.

As mentioned in the introduction, experimental indications of the quantization can be found in the work of Gurney *et al.*[6] However, it took seeveral years until the effect has been undoubtedly demonstrated and investigated.[49] Figure 10 taken from Ref. 49 shows the dispersion of the spin waves measured in two arrays of permalloy stripes of 1.8 μm width and 20 nm thickness with different separations between the stripes (full and opened symbols, correspondingly). The component of the wave vector of the spin wave along the stripe is zero, while the transversal component of the wave vector has been changed as shown in the figure. A discrete dispersion spectrum due to the quantization of the transversal component of the wave-vector described by (23) is clearly seen. Each quantized mode was observed over a continuous interval of the transferred wave vector. Its frequency remains constant while the intensity changes with the wave vector. The coincidence of the data obtained for the two arrays with different stripe separations corroborates the single-stripe nature of the observed effect. One should note that the "open" boundary condition, neglecting the dipole pinning has been used in Ref. 49 for calculation of the quantized frequencies. Therefore, the mode enumeration starts with $m = 0$ instead of the correct $m = 1$.

Figure 11 shows BLS spectra obtained for an array of 1 μm wide and 35 nm thick permalloy stripes in a longitudinal magnetic field of 500 Oe for different transversal components of the wave-vector as indicated. The measured frequencies of the quantized modes as a function of the wave vector are shown in Fig. 12. The mode with the frequency close to 20.1 GHz (shown by circles in Fig. 12) can be identified as the perpendicular standing spin wave (PSSW).[32] The modes with lower frequencies (squares) are the Damon-Eshbach modes quantized due to the confinement caused by the finite width of the stripe. Values of these frequencies calculated using (19) are shown by horizontal lines for comparison. The dashed line in Fig. 12 shows the dipole-exchange dispersion for spin waves in an unpatterned permalloy film calculated using the approach proposed.[38]

The theory described in Section 4 predicts not only the frequencies of the quantized modes (19) or (25), but their profiles as well. A comparison between the experimentally measured BLS intensities and the results of the calculation based on (2) is presented in Fig. 13. Shown are the BLS intensities from different quantized modes normalized to the intensity of the PSSW mode. Squares in Fig. 13 correspond to the experimentally measured intensities. The gray curves represent the calculated intensities using (2). It is assumed that the intensity of the PSSW modes is independent of the transferred wave vector as the total wave vector of the PSSW mode is mainly determined by its out-of-plane component (proportional to $1/L$) which is at least by a factor of ten larger than the maximum in-plane wave vector for the described measurements. The only fitting parameter used

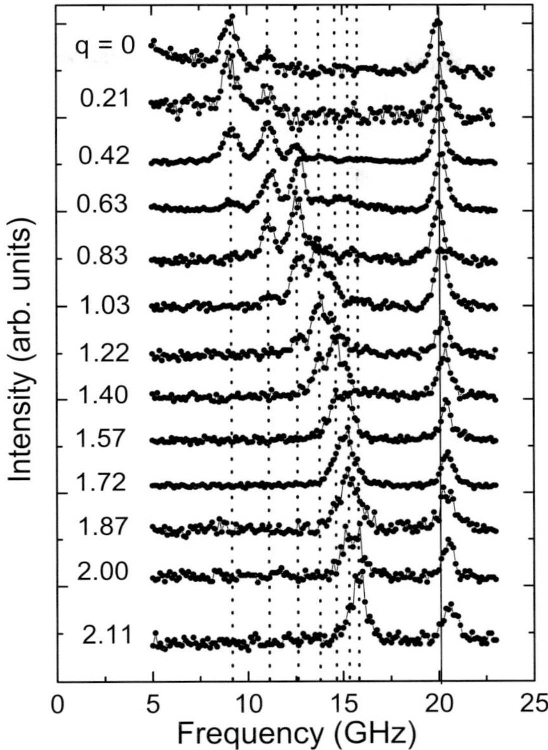

Fig. 11. BLS spectra of longitudinally magnetized 1 μm wide and 35 nm thick stripes for different transferred wave vectors in an external field of 500 Oe. The transversal component of the wave vector q is given in units of 10^5 cm^{-1}. The vertical solid line indicates the nearly constant frequency of the PSSW mode, the vertical dotted lines indicate the fixed frequencies of the quantized spin-wave modes.

in Fig. 13 is the scaling factor along the intensity axis which is the same for all modes. The mode profiles have been calculated taking into account the dipolar pinning. The very good agreement between the experimental data and the results of the calculation confirms that the observed spin-wave modes are in fact quantized Damon-Eshbach modes and justifies the proposed boundary conditions.

Figure 14 (taken from Ref. 50) addresses the BLS intensity for a given mode ($m = 4$) in more detail. Shown are the calculated intensities corresponding to three pinning conditions: complete unpinning, effective dipole pinning from (20), and complete pinning. It is evident from Fig. 14, that

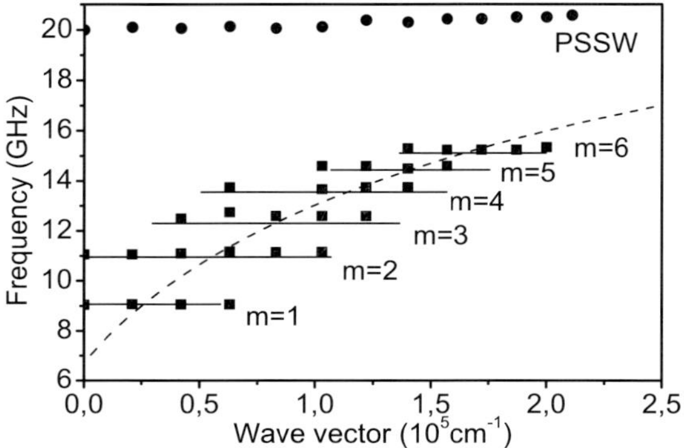

Fig. 12. Frequencies of the spin waves in longitudinally magnetized 1 μm wide and 35 nm thick stripes in an external field of 500 Oe. The squares correspond to the quantized Damon-Eshbach modes, the circles to PSSW. The horizontal solid lines are the frequencies of quantized spin-wave modes calculated using (19).

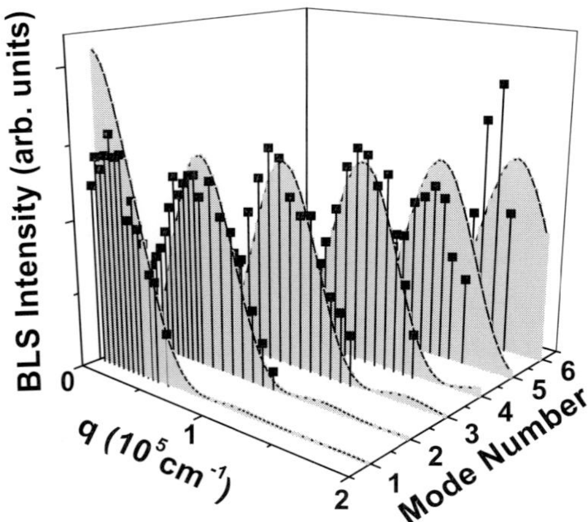

Fig. 13. Measured relative BLS-intensities for different quantized modes (black squares) as a function of the transferred wave vector q and the mode quantization number m for 1 μm wide stripes magnetized by the static magnetic field of 100 Oe in comparison to the results of calculations based on (2)(gray colored curves).

the unpinned boundary condition does not describe the experimental data. However, it is impossible to distinguish experimentally between the effective dipolar pinning and the complete pinning. This is due to a large aspect ratio of the stipe $p = \frac{L}{w} \ll 1$ causing a strong pinning.

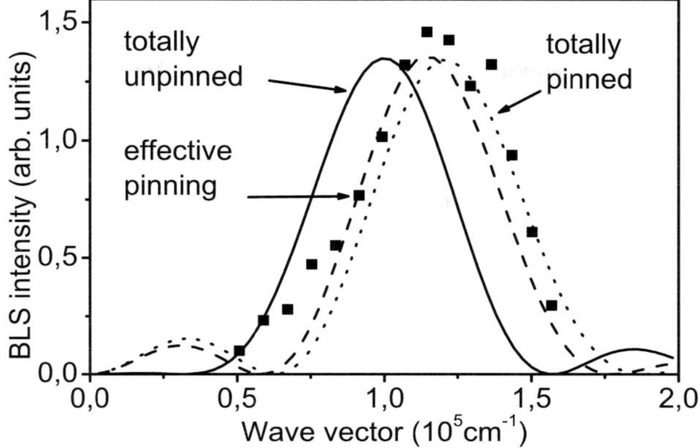

Fig. 14. BLS intensity as a function of the transferred wave vector for the mode $m = 4$. The squares indicate the experimental data while the lines correspond to the Fourier transforms of three different mode profiles: totally unpinned (full line), totally pinned (dotted line), effectively pinned (dashed line).

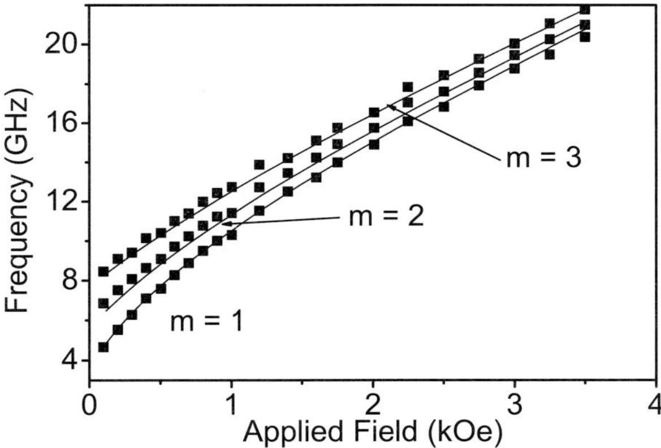

Fig. 15. Spin-wave frequency of longitudinally magnetized 1.8 μm wide and 40 nm thick stripes as a function of the applied field. The transferred wave vector is $q = 0.3 \times 10^5$ cm^{-1}.

The field dependence of the frequencies of the quantized spin-wave modes is shown in Fig. 15. The solid lines shown in the figure are calculated based on the above described approach. The figure is another demonstration of a nice agreement between the theory and the experiment.

Summarizing this section one can conclude that the effect of spin-wave quantization due to lateral confinement in longitudinally magnetized stripes is well understood and can be qualitatively described using existing models. The boundary conditions for the mode profiles are determined by the effective dipole pinning, caused by inhomogeneous dynamic dipole fields close to the edges of the stripes.

6. Spin-Wave Localization in Transversally Magnetized Stripes

If the external field is applied along the width of a stripe, both the static and the dynamic internal fields are strongly inhomogeneous due to lateral confinement. Spin waves propagating along the field are affected by confinement effects in this case. Since waves in a film with $\mathbf{q}\|\mathbf{M}_S$ are known magnetostatic backward volume waves (MSBVW),[40] the transversal experimental geometry sometimes is called the MSBVW-geometry.

It has been found that the spin-wave-well (SWW) effect can be observed in this geometry.[10] As discussed in Section 4, this effect causes a localization of spin-wave modes close to the edges of the stripe.

Figure 16(a) shows a BLS spectrum for an array of permalloy stripes in the longitudinal geometry illustrating the spin-wave quantization discussed in the previous section (Q-DE-modes). The BLS spectrum shown in Figure 16(b) was recorded for the same stripes in the transversal geometry. The low-frequency part of the spectrum contains two types of lines. The first line, indicated as "Band" is quite broad and changes its shape and position as the transferred wave vector is changed. A narrow low-frequency line indicated as LM does not change neither its shape, nor its position with the wave vector.

The high-frequency peaks of the both spectra can be easily identified as PSSW modes. Note that the PSSW peak in the transversal geometry is shifted towards the lower frequencies and demonstrates a splitting. As discussed above, the frequencies of PSSW modes are determined by the exchange interaction and the internal field H_i. Therefore the PSSW-lines can provide an information on the internal field in the stripe. From the figure we can conclude that the internal field in the transversal geometry is reduced and there are regions in the stripe with different values of the internal field, which is in agreement with the above theoretical analysis. As mentioned, in the MSBVW geometry both the static and the dynamic inter-

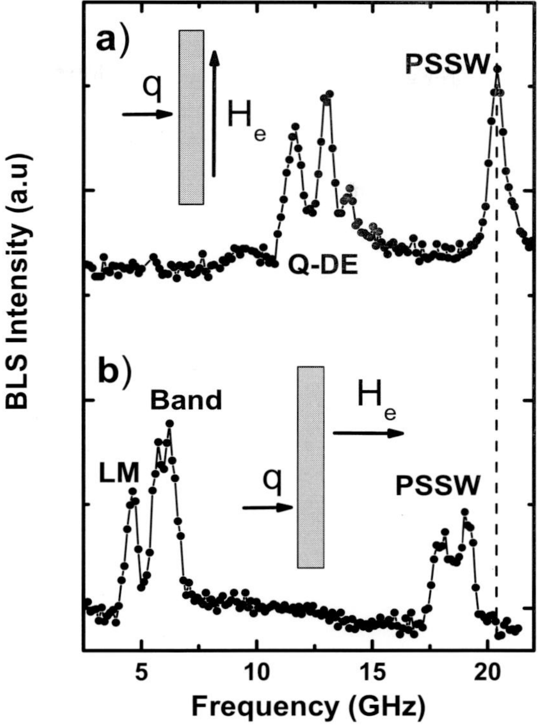

Fig. 16. BLS spectra of arrays of 33 nm thick, 1 μm wide permalloy stripes for a transferred wave vector $q = 1 \times 10^5$ cm^{-1} and an external field of $H_e = 500$ Oe for (a) the longitudinal and (b) transverse geometries, correspondingly, as indicated by the insets. Note a splitting of the PSSW-peak in the spectrum (b). Q-DE stands for the quantized spin-wave modes, LM stands for the localized mode.

nal magnetic fields in the stripe are strongly inhomogeneous. Nevertheless, for an external applied field $H_e \gg p \cdot 4\pi M_S$, where $p = L/w$ is the aspect ratio, the magnetization is parallel to the applied field over almost the entire stripe. Close to the edges of the stripe the magnetization rotates to the direction parallel to the edges in order to minimize the dipole energy. A detailed analysis reveals that such a distribution of the magnetization causes an inhomogeneous internal static field H_i having a broad maximum in the center of the stripe and vanishing near the edges of the stripe.[41] The dynamics of the spins in the stripe is determined by an effective field H_{eff}, which comprises the internal magnetic field and the exchange contribution

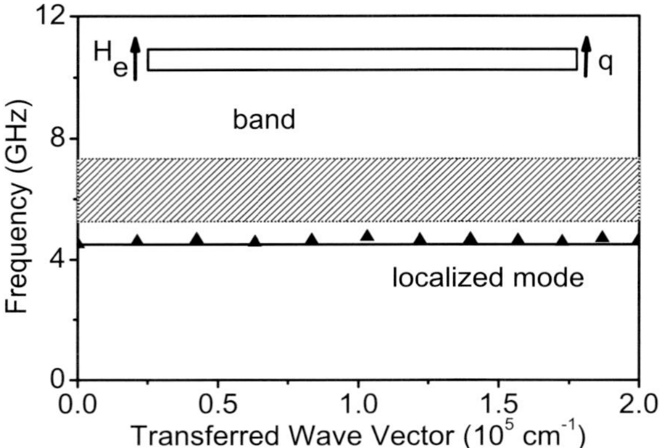

Fig. 17. Measured dispersion spectrum of a 1 μm wide stripe in an external field of 500 Oe applied along the width of the stripe. The frequency of the localized mode (full line) is calculated using (34).

$H_{\rm eff} = H_{\rm i} + H_{\rm ex}$. However, the exchange contribution is only important at the edges of the stripe where the magnetization is strongly inhomogeneous. Figure 7 shows $H_{\rm i}$ together with the distribution of the static magnetization for a 2 μm wide stripe. The shown profile explains the features of the PSSW mode observed in the transversal geometry: the internal field (i.e., the frequency of the PSSW mode) is essentially reduced with respect to longitudinal geometry, where $H_{\rm i} = H_{\rm e}$. In the saturated central region of the stripe $H_{\rm i}$ is close to its maximum value of 500 Oe. In the unsaturated regions near the stripe edges (± 1 μm) $H_{\rm i}$ is zero. As a consequence, a split PSSW-peak corresponding to two values of the frequencies is observed.

Thus, the existence of the edge regions with a complicated domain-wall-like structure can be confirmed by the analysis of the high-frequency part of the spectrum presented in Fig. 16. These regions characterized by a fast change of the internal field build the well for spin waves as discussed in Section 4. The low frequency part of the spectrum demonstrates the existence of spin-wave states localized in the well. In fact, the stronger broad peak between 5 and 7 GHz corresponds to a band of non-localized spin-wave modes (see Section 4). The peak with the lower frequency is a localized spin-wave mode corresponding to a spin-wave state in the well. The strong spatial localization of the mode is illustrated by its dispersion shown in Fig. 17. As seen in the figure, the frequency of the mode does not depend on the transferred wave vector. On the other hand, the mode

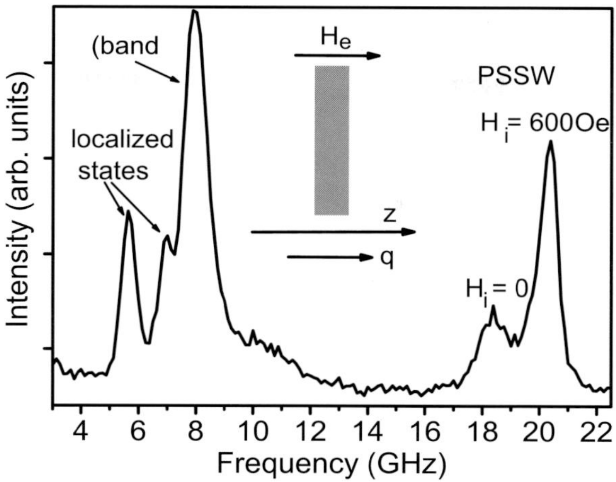

Fig. 18. BLS spectrum of a transversally magnetized 1 μm wide stripe for a transferred wave vector $q = 0.47 \times 10^5$ cm^{-1} and an external field of $H_\mathrm{e} = 800$ Oe.

is observed over the entire interval of the accessible transferred wave vector and its BLS intensity shows no significant change as a function of the wave vector. Using the relation between the profile of the dynamic magnetization in the real space and the BLS intensity profiles in the $q-$space via the Fourier transformation discussed above one can easily estimate the upper limit for the localization width of the mode: $\Delta r = \pi/q_{max} \simeq 120$ nm.

It is interesting to track the evolution of the observed modes with changing external applied field, H_e, which is illustrated in Figs. 18 and 19 (taken from Ref. 12). Since the maximum value of H_i increases with increasing H_e and the minimum value $H_\mathrm{i} = 0$, the inhomogeneity of H_i becomes more profound with growing H_e. Indeed, comparing the spectra shown in Figs. 16 and 18, which are recorded at $H_\mathrm{e} = 500$ and 800 Oe, respectively, one can see that the number of the localized modes increases with increasing field. Fig. 19 shows the mode frequencies as a function of the applied external field H_e. While the magnitude of H_e is small and not sufficient to create a region of saturated magnetization in the stripe, only a single PSSW mode corresponding to $H_\mathrm{i} = 0$ is present in the spectrum. At some critical field $H_\mathrm{e} = H^*$ the PSSW peak begins to split. The first peak corresponding to $H_\mathrm{i} = 0$ retains its frequency, but looses its intensity with increasing H_e (compare Figs. 16 and 18). This fact confirms the expectation that the regions with $H_\mathrm{i} = 0$ shrink with increasing H_e. The frequency of the

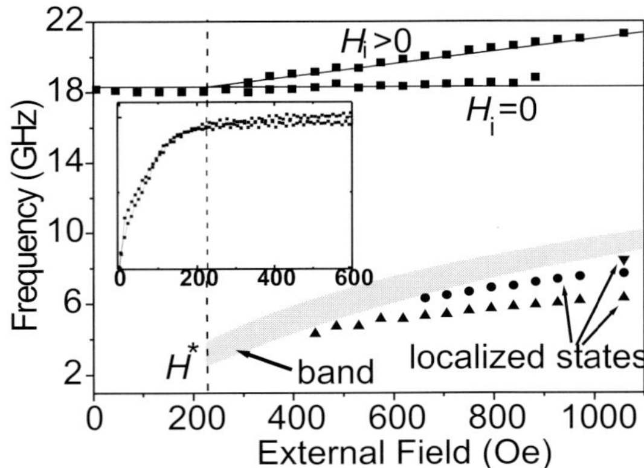

Fig. 19. Frequencies of the modes observed in a transversally magnetized 1 μm wide stripe at $q = 0.47 \times 10^5$ cm^{-1} as a function of H_e. The inset shows the magnetization curve of the stripe for reference. The vertical dashed line marks the critical field H^* both in the main figure and in the inset. Note the constant frequency of 18 GHz of one of the PSSW modes and the increase in frequency for the other mode for $H_e > H^*$. The gray region illustrates the band of unresolved excitations.

second peak (which appears at $H_e = H^*$) increases with the applied field, indicating the increase of H_i in the central part of the stripe. The observed value of the critical field $H^* = 220$ Oe is in agreement with the calculated demagnetizing field in the stripe center, $H_d(z = 0)$, based on (27) and (28), and with static magnetization curves, shown as the inset in Fig. 19.

One can understand the evolution of the localized modes with increasing H_e keeping the analogy with the quantum mechanical potential well. More localized modes fit into a deeper well. Therefore the character of a given mode changes with the increase of the applied field. A mode which is not localized at a low field can become localized with increasing field.

7. Spin-Wave Modes in Rectangular Platelets

If a magnetic element confined in all dimensions is considered, there is no orientation of the external field which provides a homogeneous static internal magnetic field. However, as shown above, the magnetic dynamics in a rectangular platelet can be treated with a good accuracy as a combination of the two stripe geometries considered above, provided the thickness of the platelet is much smaller than its lateral dimensions. Let us assume that the applied field is directed along one of the edges of the platelet. If the

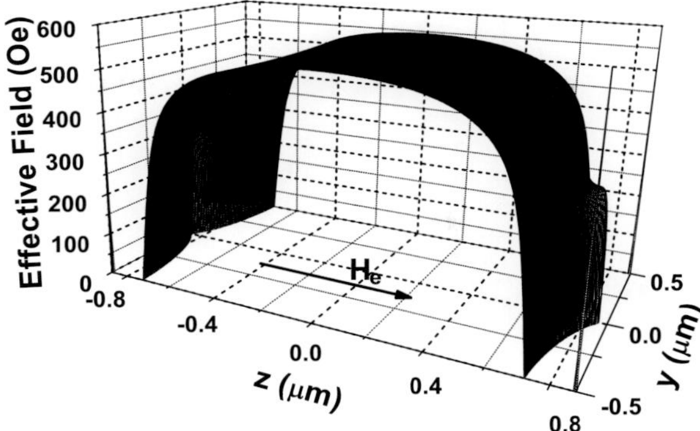

Fig. 20. The z-component of the effective field in a $1 \times 1.75\ \mu\text{m}^2$ rectangular platelet of the thickness $L = 35$ nm for an applied field of 600 Oe calculated with OOMMF.

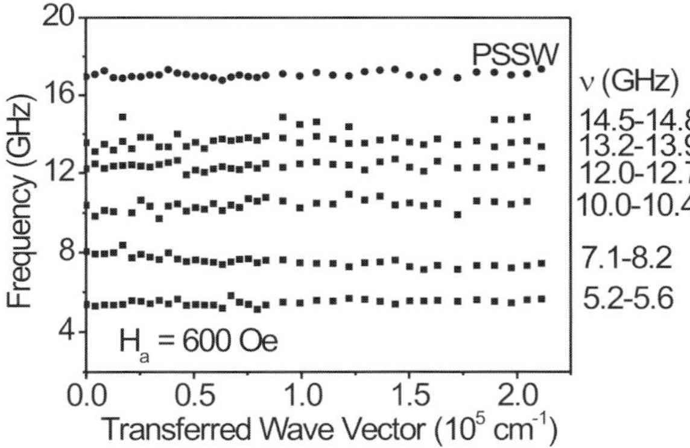

Fig. 21. Dispersion spectrum of the $1 \times 1.75\ \mu\text{m}^2$ rectangular platelet in a field of 600 Oe applied along its the long side. The transferred wave vector is oriented along the long axis of the platelet as well.

field is large enough, the magnetization is essentially homogeneous and parallel to the applied field. Nevertheless, the effective internal field is strongly inhomogeneous as shown in Fig. 20. The field has been calculated using the OOMMF code for micromagnetic calculations.[43]It is seen that the inhomogeneity occurs mainly along the direction of the applied field (z-direction). The inhomogeneity along the y-direction is not so pronounced.

Figure 21 (from Ref. 51) showing the dispersion spectrum in a rectangular platelet illustrates the spin-wave quantization. The analysis of the

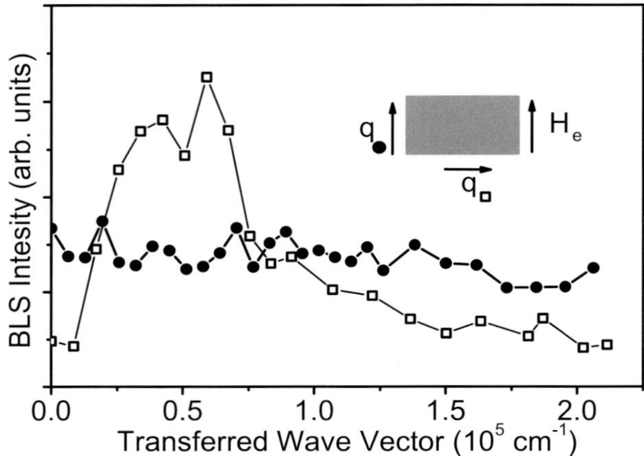

Fig. 22. BLS intensity for a spin-wave mode for rectangular platelets as a function of the transferred wave vector for two orientations of the latter.

spin-wave dynamics in such an element, is more complicated as that for the stripes. However, one concludes that the observed frequency splitting is due to spin-wave quantization along the direction which is perpendicular to the applied magnetic field. A two-dimensional "Fourier-mapping" of the spin-wave modes can bring more information on the mode profiles. For this purpose the BLS intensity for a given mode is recorded as a function of the transferred wave vector along two lateral directions. As an example of such an analysis, Fig. 22 (from Ref. 51) shows the BLS intensity of a mode as a function of the absolute value of the transferred wave vector parallel and perpendicular to the applied field. As can be seen, two curves exhibit very different dependencies. For a transferred wave vector parallel to the applied field an almost constant BLS intensity over a wide interval of wave vectors indicates a strong localization of the mode in real space. Contrarily, for a transferred wave vector perpendicular to the applied field the change in intensity confirms that the mode should have a sin-profile characteristic for quantized modes.

8. Nonlinear Spin-Wave Modes in Confined Geometries

In the previous Sections we considered the small-amplitude spin-wave eigenmodes, for which the condition $m \ll M_S$ is satisfied. As discussed in Section 4, in this approximation the Landau-Lifshitz torque equation (3) can be linearized with respect to the amplitude of the dynamic magnetization, which significantly simplifies the theoretical analysis. Nevertheless, for many processes in confined magnetic structures such as, e.g., switching or microwave emission by spin-torque-transfer nano-oscillators, the amplitudes of dynamic magnetization are essentially large[52] and this linear approximation cannot be used.

It is evident that experimental addressing of nonlinear magnetization dynamics directly on the microscopic scale is rather difficult. On the other hand, there are techniques for investigation of nonlinear magnetic dynamics at large scales. Therefore, one can try to scale the problem to large spatial scales easily accessible experimentally. Strictly speaking, the magnetic dynamics defined by both the exchange and the magnetic dipole interactions is not spatially scalable, since these interactions define the exchange-dipole correlation length $\xi = \sqrt{2\pi M_S{}^2/A}$, where A is the exchange stiffness constant. However, since the correlation lengths for 3d-metals are 5-50 nm, the problems with typical sizes much larger than ξ can be scaled upwards. The scaling is still not trivial, since the mean free path of spin waves has to be scaled appropriately, which is not easy for metals. Using low-loss dielectric magnetic films of yttrium iron garnet (YIG),[27] one can solve this problem as well. YIG films allow one to realize confined magnetic structures with lateral dimensions in the millimeter range and with the ratio between the lateral size and the thickness being very close to that for microscopic metallic magnetic elements. Spin waves in such elements can now be studied with the help of spatially resolved BLS technique and the dynamic magnetization in the linear and nonlinear eigenmodes can be directly visualized.

Examples of the distributions of the dynamic magnetization mapped using the above experimental technique are shown in Fig. 23 (from Ref. 53) for YIG squares with the side length of $w = 2$ mm. The figure presents two-dimensional maps of the spin-wave intensity (i.e., the dynamic magnetization squared) and their spatial Fourier spectra for the spin-wave eigenmodes (2,1) and (4,1). The notation of the modes is connected to the number of the antinodes in the spatial distribution of the dynamic magnetization in the mode: e.g., mode (2,1) has one antinode in the direction perpendicular to the applied magnetic field $\mathbf{H}_e$ and two antinodes in the direction parallel to $\mathbf{H}_e$. The data are given for the linear and the strongly nonlinear cases differing in the power of the microwave signal used for the excitation of the dynamic magnetization in the squares. As seen from Fig. 23, the distributions of the dynamic magnetization of the eigenmodes in the linear case

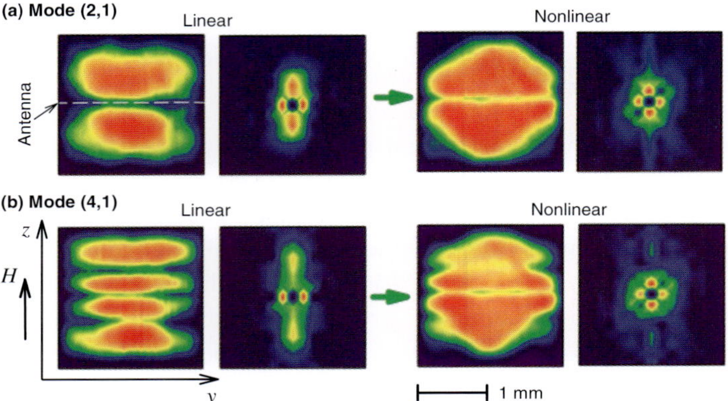

Fig. 23. Measured maps of the dynamic magnetization and their two-dimensional Fourier spectra for the eigenmodes (2,1) (a) and (4,1) (b) in the linear and strongly nonlinear regimes.

demonstrate clearly defined standing waves in two directions, which is in accordance with the theoretical concept of Section 4. As the amplitude of the dynamic magnetization increases and comes into the nonlinear range, the response of the system becomes essentially nonlinear. The concept of factorization of the two-dimensional magnetization distribution into two one-dimensional spin-wave eigenfunctions does not work well under these conditions. Spin-wave eigenmodes exhibit significant changes, which can be described as a widening and merging of the neighboring maxima of the standing waves, resulting in a flattening of the standing-wave profile. The analysis of the spatial Fourier spectra reflecting the distribution of the dynamic magnetization in the wave-vector space shows that the nonlinearity leads to a decrease of the spatial periodicity of the eigenmodes: the peaks corresponding to the wave vectors $k_z = 2\pi/w$ and $4\pi/w$ clearly seen in the linear case, disappear in the nonlinear case, and the peaks corresponding to $k_z = \pi/w$ arise.

The nonlinear spin-wave eigenmodes cannot be understood based on the linearized Landau-Lifshitz equation used in Section 4. Instead, one has to use the nonlinear complex Ginzburg-Landau equation adapted for the description of magnetization dynamics in ferromagnetic film structures.[53] Based on this equation one can numerically simulate distributions of the dynamic magnetization of the eigenmodes in both linear and the nonlinear regimes and find out what nonlinear processes are responsible for the non-

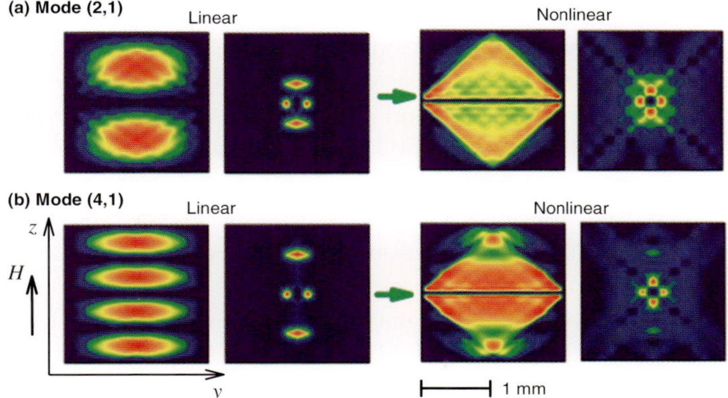

Fig. 24. Calculated maps of the dynamic magnetization and their two-dimensional Fourier spectra for the eigenmodes (2,1) (a) and (4,1) (b) in the linear and strongly nonlinear regimes.

linear modifications observed in the experiment. Figure 24 (from Ref. 53) shows the results of the simulations based on the above model. As seen from the figure, all the experimental findings are accurately reproduced within the theory. Moreover, the analysis of the relation between the parameters entering the Ginzburg-Landau equation and the results of simulations clearly shows that the dominating nonlinear effect in the system is the so-called nonlinear damping.[54] Presence of the nonlinear damping leads to a nonlinear coupling between the eigenmodes resulting in the energy channeling from the higher-order eigenmodes to the lower-order ones, which explains the observed nonlinear transformations of the spatial Fourier spectra.

Note that, even though the above experimental investigations were performed for millimeter-size squares, the made conclusions should also be valid for nano-scale magnetic square islands. It is also worth noting that the phenomenon of nonlinear mode coupling has potential for technical applications. In particular, it can be used in practice for nonlinear amplification and/or reduction of magnetic losses of the eigenmodes.

9. Conclusions

In this Chapter we have considered the concept of lateral spin-wave confinement in different model thin film magnetic elements. We have shown that the confinement has two important aspects: (i) geometrical confinement; (ii) effects of inhomogeneous demagnetizing fields. One-dimensional confinement in stripes magnetized along their length corresponds to the simplest situation, since the second aspect of the confinement can be neglected in this case. A rather transparent spin-wave quantization effect, caused by the geometrical quantization of the wave vector, is observed. In stripes magnetized along their widths both aspects must be considered, and the effect of spin-wave localization in the wells created by the inhomogeneous demagnetizing field appears. Confined two-dimensional spin-wave modes in rectangular platelets combine the spin-wave properties of both above one-dimensional geometries. The next level of complexity represents the nonlinear case, where the nonlinear two-dimensional modes cannot be obtained as a combination of one-dimensional modes.

Acknowledgements

The authors are deeply indebted to their colleagues who contributed to the original works making the basis for this review. Financial support by the Deutsche Forschungsgemeinschaft is gratefully acknowledged.

References

1. B. Hillebrands, K. Ounadjela (Eds.): *Spin Dynamics in Confined Magnetic Structures I*, 1st edn. (Springer, Berlin, Heidelberg, New York, 2002). B. Hillebrands, K. Ounadjela (Eds.): *Spin Dynamics in Confined Magnetic Structures II*, 1st edn. (Springer, Berlin, Heidelberg, New York, 2003).
2. M. L. Plumer, J. van Ek, D. Weller (Eds.): *The Physics of Ultra-High-Density Magnetic Recording*, 1st edn. (Springer, Berlin Heidelberg New York, 2001).
3. N. Smith, P. Arnett, *Appl. Phys. Lett.* **78**, 1448 (2001).
4. J.-G. Zhu, *J. Appl. Phys.* **91**, 7273 (2002).
5. Y. Zhou, A. Roesler, J.-G. Zhu, *J. Appl. Phys.* **91**, 7276 (2002).
6. B. A. Gurney, P. Baumgart, V. S. Speriosu, R. E. Fontana, A. Patlac, A. Logan, P. Humbert, Proc. of the 13th Int'l Colloq. on Mag. Thin Films and Surfaces, Glasgow **P7.12**, 474 (1991).
7. C. Kittel, *Phys. Rev.* **73**, 2 (1948).
8. L.R. Walker, *Phys. Rev.* **105**, 2 (1957).
9. R. I. Joseph, E. Schlömann, *J. Appl. Phys.* **36**, 1579 (1965).
10. J. Jorzick, S. O. Demokritov, B. Hillebrands, D. Berkov, N. L. Gorn, K. Guslienko, A. N. Slavin, *Phys. Rev. Lett.* **88**, 047204 (2002).
11. K. Yu. Guslienko, S. O. Demokritov, B. Hillebrands, A. N. Slavin, *Phys. Rev. B* **66**, 132402 (2002).

12. C. Bayer, S. O. Demokritov, B. Hillebrands, A. N. Slavin, *Appl. Phys. Lett.* **82**, 607 (2003).
13. J. P. Park, P. Eames, D. M. Engebretson, J. Berezovsky, P. A. Crowell, *Phys. Rev. Lett.* **89**, 277201 (2002).
14. D. D. Awschalom and D. P. DiVeincenzo, *Phys. Today*, 48:43, 1995.
15. M. Hehn, K. Ounadjela, J. P. Bucher, F. Rousseaux, D. Decanini, B. Bartenlian, and C. Chappert, *Science*, 272:1782, 1996.
16. C. Chappert, H. Bernas, J. Ferre, V. Kottler, J.-P. Jamet, Y. Chenm, E. Cambril, T. Devolder, F. Rousseaux, V. Mathet, and H. Launois, *Science*, 280:1919, 1998.
17. W. Wernsdorfer, K. Hasselbach, D. Mailly, B. Barbara, A. Benoit, L. Thomas, and G. Suran, *J. Magn. Magn. Mater.*, 145:33, 1995.
18. V. E. Demidov, S. O. Demokritov, B. Hillebrands, M. Laufenberg, P. P. Freitas, *Appl. Phys. Lett.* **85** (2004) 2866.
19. M. Grimsditch, Y. Jaccard, and I. K. Schuller, *Phys. Rev. B* **58**, 11539 (1998).
20. Z. K. Wang, M. H. Kuok, S. C. Ng, D. J. Lockwood, M. G. Cottam, K. Nielsch, R. B. Wehrspohn, U. Gösele, *Phys. Rev. Lett.* **89**, 027201 (2002).
21. B. Hillebrands, C. Mathieu, M. Bauer, S. O. Demokritov, B. Bartenlian, C. Chappert, D. Decanini, F. Rousseaux, and F. Carcenac, *J. Appl. Phys.* **81**, 4993 (1997).
22. S. M. Cherif, C. Dugautier, J. F. Hennequin, and P. Moch, *J. Magn. Magn. Mater.* **175**, 228 (1997).
23. A. Ercole, A. O. Adeyeye, J. A. C. Bland, and D. G. Hasko, *Phys. Rev. B* **58**, 345 (1998).
24. T. Aign, P. Meyer, S. Lemerle, J. P. Jamet, J. Ferré, V. Mathet, C. C. Gierak, J. Gierak, C. Vieu, F. Rousseaux, H. Laumois, and H. Bernas, *Phys. Rev. Lett.* **81**, 5656 (1998).
25. S. M. Cherif, Y. Roussigne, P. Moch, J. F. Hennequin, and M. Labrune, *J. Appl. Phys.* **85**, 5477 (1999).
26. Y. Roussigné, S. M. Chérif, C. Dugautier, P. Moch, *Phys. Rev. B* **63**, 134429 (2001)
27. V. E. Demidov, U.-H. Hansen, and S.O. Demokritov, *Phys. Rev. Lett.* **98** (2007) 157203
28. G. Gubbiotti, M. Kostylev, N. Sergeeva, M. Conti, G. Carlotti, T. Ono, A. Stashkevich, *Phys. Rev. B* **70**, 224422 (2004)
29. S. I. Kiselev, J. C. Sankey, I. N. Krivorotov, N. C. Emley, R. J. Schoelkopf, R. A. Buhrman, D. C. Ralph, *Nature* **425**, 380 (2003).
30. W. H. Rippard, M. R. Pufall, S. Kaka, S. E. Russek, and T. J. Silva, *Phys. Rev. Lett.* **92**, 027201 (2004).
31. S. O. Demokritov, B. Hillebrands, and A. N. Slavin, *Phys. Rep.*, 348:441, 2001.
32. J. Jorzick, S. O. Demokritov, C. Mathieu, B. Hillebrands, B. Bartenlian, C. Chappert, F. Rousseaux, A. N. Slavin, *Phys. Rev. B* **60**, 15194 (1999).
33. J. R. Sandercock, *Trends in Brillouin scattering: Studies of opaque materials, supported films, and central modes, in: Light Scattering in Solids III*

(M. Cardona and G. Güntherodt (eds.), Springer Verlag, Berlin, Heidelberg, New York, 1982).
34. B. Hillebrands, *Rev. Sci. Instrum.* **70**, 1589 (1999)
35. L. Landau and E. Lifshitz, *Phys. Z. Sowjetun.*, 8:153, 1935.
36. A. G. Gurevich and G. A. Melkov, *Magnetization Oscillations and Waves* (CRC Press, New York, 1996).
37. V.Lvov, *Wave Turbulence Under Parametric Excitation* (Springer, 1994).
38. B. A. Kalinikos, A. N. Slavin, *J. Phys. C* **19**, 7013 (1986).
39. C. Bayer, J. Jorzick, B. Hillebrands, S.O. Demokritov, A. N. Slavin, K. Guslienko, D. Berkov, N. Gorn, M. P. Kostylev, in *Spin Dynamics in Confined Magnetic Structures*, Eds. B. Hillebrands, A. Thiaville, *Topics Appl. Phys.* (Springer, Berlin, 2006).
40. R. W. Damon, J. R. Eshbach, *J. Phys. Chem. Solids* **19**, 308 (1961).
41. P. Bryant, H. Suhl, *Appl. Phys. Lett.* **54**, 2224 (1989).
42. Ernst Schlömann, *J. Appl. Phys.* **35**, 159 (1964).
43. M. J. Donahue, D. G. Porter, *OOMMF User's Guide, Version 1.0,* in Interagency Report No. NISTIR 6376, National Institute of Standards and Technology, Gaithersburg, Maryland, 1999.
44. C. Bayer, J. P. Park, H. Wang, M. Yan, C. E. Campbell, P.A. Crowel, *Phys. Rev. B* **69**, 134401 (2004).
45. M. Bailleul, D. Olligs, C. Fermon, *Phys. Rev. Lett.* **91**, 137204 (2003).
46. Y. Roussigné, S.M. Chérif, P. Moch, *J. Magn. Magn. Mater.* **263**, 289 (2003).
47. Y. Roussigné, S. M. Chérif, P. Moch, *J. Magn. Magn. Mater.* **268**, 89 (2004).
48. G. Gubbiotti, M. Conti, G. Carlotti, P. Candeloro, E. Di Fabrizio, K. Yu. Guslienko, A. Andre, C. Bayer, A. N. Slavin, *J. Phys.: Condensed Matter* **16**, 7709-7721 (2004)
49. C. Mathieu, J. Jorzick, A. Frank, S. O. Demokritov, B. Hillebrands, A. N. Slavin, B. Bartenlian, C. Chappert, D. Decanini, F. Rousseaux, E. Cambril, *Phys. Rev. Lett.* **81**, 3968 (1998).
50. S. O. Demokritov, *J. Phys.: Condens. Matter* **15**, S2575 (2003).
51. J. Jorzick, C. Krämer, S. O. Demokritov, B. Hillebrands, B. Bartenlian, C. Chappert, D. Decanini, F. Rousseaux, E. Cambril, E. Søndergard, M. Bailleul, C. Fermon, A. N. Slavin, *J. Appl. Phys.* **89**, 7091 (2001)
52. J. C. Sankey, I. N. Krivorotov, S. I. Kiselev, P. M. Braganca, N. C. Emley, R. A. Buhrman, and D. C. Ralph, *Phys. Rev. B* **72** (2005) 224427.
53. M. P. Kostylev, V. E. Demidov, U.-H. Hansen, and S. O. Demokritov, *Phys. Rev. B* **76** (2007) 224414.
54. M. M. Scott, C. E. Patton, M. P. Kostylev, and B. A. Kalinikos, *J. Appl. Phys.* **95**, 6294 (2004).

Chapter 2

BRILLOUIN LIGHT SCATTERING STUDY OF SPIN DYNAMICS IN PATTERNED NANO-ELEMENTS: FROM SINGLE-LAYER TO MULTILAYERED STRUCTURES

G. Gubbiotti, M. Madami, S. Tacchi, and G. Carlotti
CNISM e Dipartimento di Fisica, Università di Perugia,
Via A Pascoli, 06123 Perugia, Italy
E-mail: Gubbiotti@fisica.unipg.it

A Brillouin light scattering experimental investigation of the high frequency dynamical properties of multi-layered patterned elements with sub-micrometric lateral dimensions is reported. The different magnetic excitation spectra of the ferromagnetic layers employed in such structures, combined with the effects of interlayer coupling and finite lateral dimensions, make the dynamics of such elements different from that measured in a single-layer magnetic film. The effect of lateral confinement in one and two in-plane directions is explored in patterned samples made of stripes with infinite length, as well as rectangular and elliptical elements, respectively.

1. Introduction

The study of spin dynamics in patterned magnetic structures is an interesting and challenging research field. Its development started about a decade ago when advances in lithography have made possible the realization of patterns of nanometric magnetic elements with controlled lateral shape and dimensions. New properties of spin excitations have been observed, such as the existence of quantized non-dispersive resonant modes and localized excitations due to the lateral confinement and to the inhomogeneity of the internal magnetic field, respectively.[1] It has been recently shown that dynamical modes play an important role also in assisting and driving the magnetization reversal process. When an abrupt magnetization switching occurs, it is accompanied by a soft magnetic mode (its frequency goes to zero) whose spatial symmetry determines the initial steps (onset) of the microscopic reversal path.[2-3] These new properties were firstly observed by Brillouin light scattering (BLS) which has proven to be a very efficient technique for directly measuring dispersion characteristics of magnons on laterally confined nanostructures.[4-8] In a BLS experiment, a beam of monochromatic light is focused onto the surface of a sample and the back-scattered light is analysed in frequency by a Fabry-Perot interferometer. It is noteworthy that in standard experiments the area of the focused laser spot has a diameter of about 20–30 microns and therefore several thousands of nano-elements are usually illuminated, so the information obtained is averaged over such a large number of elements. While most of the studies have dealt with single-layer magnetic elements, with different shape and ground state, relatively far apart from the other elements to avoid coupling, there has been a growing interest in understanding the properties of magnetic normal modes in dipolarly coupled magnetic elements. When the inter-element separation is sufficiently small (typically below 100 nm) dipolar coupling leads to the formation of collective modes propagating through the array, with the appearance of Brillouin zones determined by the artificial periodicity. Detailed investigation of such an effect is, however, a quite complicated task because of the technical difficulty to produce arrays of magnetic elements with a very narrow distribution of shapes, sizes and distances. The statistical variations of these parameters can often hide, at least partially, the effect of magnetostatic interaction in arrays of thousands elements. In addition, to quantitatively interpret the effect of magnetostatic coupling, micromagnetic simulations are necessary and these are limited by the complexity in treating the demagnetizing energy over a large number of elements.

A different approach to investigate the effect of dipolar coupling is to fabricate laterally confined elements composed of alternating magnetic and non-magnetic layers, as for instance in pseudo spin-valve elements. The dynamics of such elements is expected to be different from that measured in a single-layer magnetic film, due to both interlayer coupling and different magnetic excitation spectra of each layer. The principal effect of the inter-layer dipolar coupling is the formation of either "acoustic" (in-phase) or "optic" (in anti-phase) coupled spin wave modes.[9-15] The applied aspect of these studies should not be underestimated, either. Multilayer stacks have been proposed as basic units in the magnetic random access memory (MRAM) technology and are of current interest also in spintronics since, when an ac current is injected into the structure, spin waves can be excited through the so-called spin-transfer torque (STT) effect.[16-18] Detailed knowledge of the spin-wave spectrum of such structures, including the spatial distribution of magnetic excitations, can be of great help to interpret the results of STT experiments.

The intent of this chapter is to review some relevant aspects of the experimental activity of our research group at the Department of Physics of the University of Perugia. We exploited Brillouin light scattering to gain information on thermally excited spin waves in arrays of patterned nano-elements. This work was possible thanks to fruitful collaboration with several research groups around the world. Among the other, we cite here the groups of Prof. Teruo Ono (*Institute for Chemical Research of the Kyoto University, Japan*) and of Prof. A. O. Adeyeye (*Information Storage Materials Laboratory of the, National University of Singapore, Singapore*) who fabricated the patterned samples we have studied. The interpretation of the BLS experiment was performed in collaboration with several theoreticians, such as Prof. F. Nizzoli and his co-workers (*Dipartimento di Fisica, Università di Ferrara, Italy*), Dr. M. Kostylev (*School of Physics, University of Western Australia*), Dr. Y. Roussigné (*CNRS, Laboratoire PMTM, Université Paris 13,France*) and Prof. A. N. Slavin (*Department of Physics, Oakland University,USA*).

This chapter has the following organization: after this introduction, the experimental procedure of the BLS measurements is briefly presented and the different scattering geometries are outlined in Section 2. In Sections 3 and 4 the results obtained for guided modes in isolated ferromagnetic stripes and the collective spin modes in dipolarly coupled alternating width stripes array, are presented, respectively. Section 5 addresses the micromagnetic method we used to calculate the eigen-mode frequencies and the spatial profiles in laterally confined (single- and multi-layered) structures. In Section 6 the effect of patterning on the eigen-mode spectrum of layered structures is discussed. We start from the spin wave properties of continuous ferromagnetic films and trilayers. Then we show the effect of lateral confinement in one and two in-plane directions, looking at stripes and rectangular elements, respectively. The discussion is then extended in Section 7 to the case of NiFe/Cu/NiFe dots with either circular or elliptical cross-section.

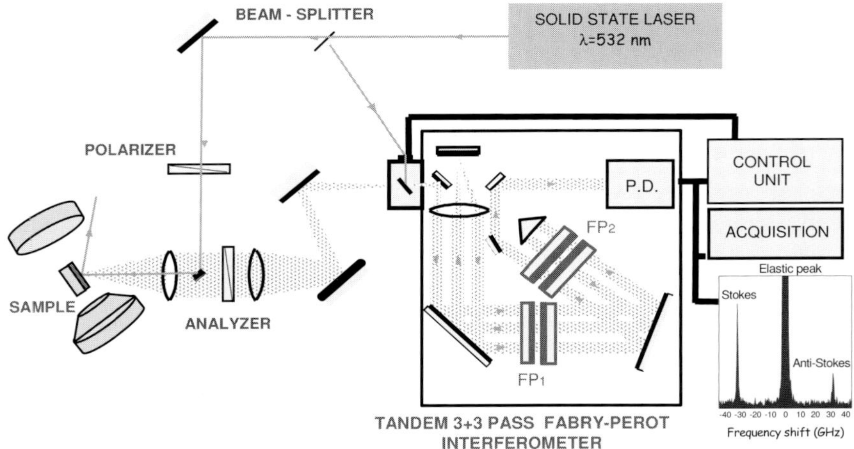

Fig. 1. Schematic representation of the BLS apparatus.

2. Experiment

The Brillouin light scattering (BLS) experiments reviewed here (Fig. 1) were carried out at the GHOST laboratory,[19] Perugia University, using a Sandercock (3+3)–pass tandem Fabry-Perot interferometer.[20] A beam of typically 200 mW of P-polarized light from a single-mode solid state laser operating at $\lambda = 532$ nm is focused on the sample surface using a camera objective of f-number 2 and focal length 50 mm and the scattered light is collected by the same objective used for focusing the laser beam (backscattering configuration). Since the light scattered from spin waves has a polarization rotated by 90° with respect to the incident one, an analyzer set at extinction (with respect to the polarization of the laser) suppresses the signal from both elastically-scattered and surface phonons-scattered light.

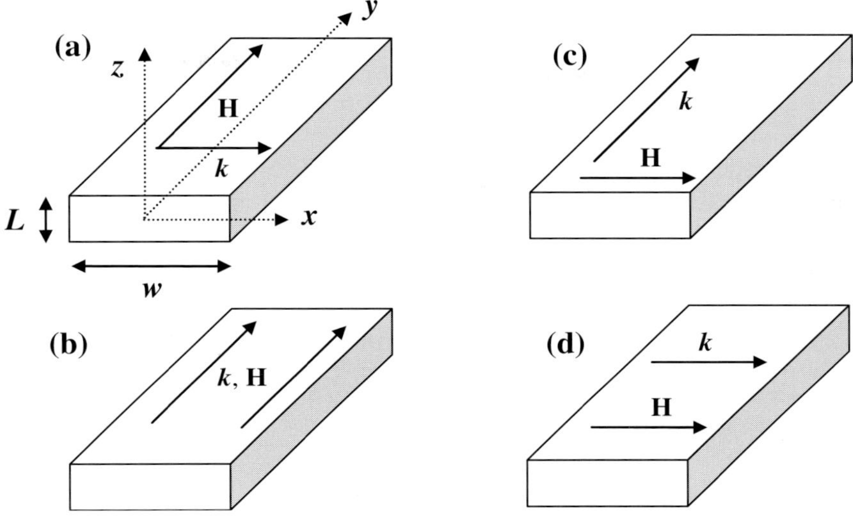

Fig. 2. Different geometries used in the Brillouin light scattering measurements. In configurations (a) and (d) the transferred wave vector is parallel to the stripes width while it is directed along the stripes length in cases (b) and (c).

The sample is mounted on a goniometer to allow rotation around the field direction, i.e. to vary the incidence angle of light, θ, between 8° and 70°. The amplitude of the wave vector entering in the scattering process is related to the angle of incidence, θ, by the following relation $k = (4\pi/\lambda)\sin\theta$, so that the magnon wave vector ranges in the interval between $0.33 \cdot 10^5$ cm^{-1} to $2.2 \cdot 10^5$ cm^{-1}. In the case of elongated elements, such as stripes, measurements have been performed in four different scattering geometries, shown in Fig.2, corresponding to different relative orientations of the external field **H**, applied parallel to the sample surface, and the scattering plane containing the transferred wave vector *k*. We note that in cases (a) and (c) the magnetic filed is orthogonal to the transferred wave vector, i.e. to the scattering plane (*x*-*z* plane and *y*-*z* plane, respectively). In contrast, for geometries (b) and (d), both the external applied field and the scattering plane (*y*-*z* plane and *x*-*z* plane, respectively) are parallel.

3. Resonating and Guided Spin Waves in Isolated Ferromagnetic Stripes

BLS studies performed on ordered arrays of magnetic stripes (or wires), defined as a very long magnetic rod with rectangular cross-section and with submicrometric width, served as a model system to shed light on the new properties of spin wave excitations in laterally confined magnetic elements. Here we report the results relative to an array of Permalloy ($Ni_{81}Fe_{19}$) stripes, fabricated on thermally oxidized Si substrates by means of e-beam lithography, electron-gun deposition, and lift-off process, having thickness L = 30 nm, width w = 600 nm and edge-to-edge spacing d = 500 nm. For this spacing each stripe can be considered as an isolate (non-interacting) magnetic element.

We present first, the results obtained in the scattering geometries of Figs. 2(a) and 2(d), i.e. with the transferred wave vector perpendicular to the stripes, because these are the scattering configurations usually employed in BLS experiments to probe the discrete spin wave resonances.[21] We recall that these configurations correspond, in the case of in-plane magnetized films, to the propagation of magnetostatic surface wave (Damon-Eshbach mode) and backward volume modes characterized by a positive and negative group velocity of the spin waves, respectively.

For stripes magnetized along their length (easy magnetization direction, as in Fig. 2 (a)), the demagnetizing field is negligible and the static magnetic field inside the stripe is spatially homogeneous. Spin waves are reflected back and fourth from the stripes edges, corresponding to stationary waves (resonances) with quantized wave numbers. Typical BLS spectra present several discrete peaks whose frequencies do not change as a function of wavevector k_x, as shown in Fig. 3(a). When the external field is applied perpendicular to the stripes (hard magnetization direction, as in Fig. 2(d)), the internal field is inhomogeneous due to the demagnetization effect which leads to the decrease of the static internal magnetic field near the stripe edges. Even in this case the modes are non-dispersive and the spectrum consists of an infinite set of standing wave resonances across the whole width of the stripe and a limited number of localized modes at lower frequency, as shown in Fig. 3(d). It has been shown in Ref. 1 that these modes, being the lowest in frequency in the spectrum, are nevertheless dominated by exchange interaction and are localized in regions of strong inhomogeneity of the internal magnetic field near the stripe edges (for this reason they are usually called edge-modes or end-modes). Together with the experimental frequencies, the results of calculation (solid lines), based on an analytical model described in Ref. 22 are shown in Figs. 3(a) and 3(d).

Although the above discussed geometries (a) and (d) are the most usual ones, we have recently performed systematic BLS measurements with the incidence plane, and thus the spin-wave wave vector, parallel to the stripes, i.e. in the geometric configurations of Figs. 2(b) and 2(c).

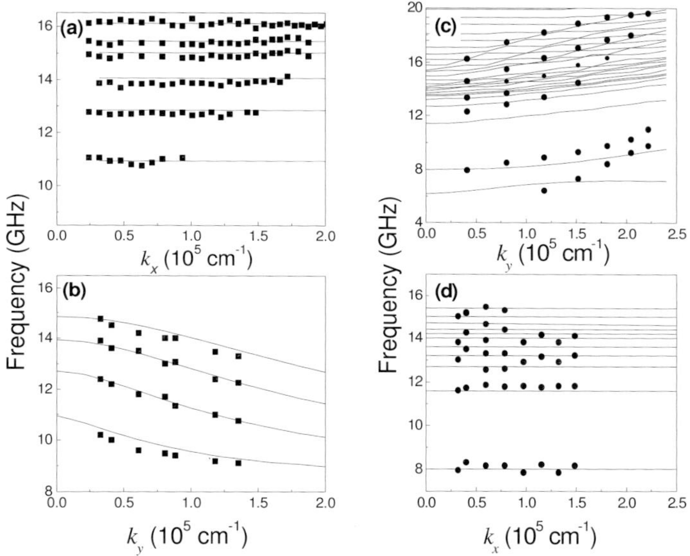

Fig. 3. Spin wave frequencies measured in the four scattering geometries shown in Fig. 2. Continuous curves are calculated frequencies using the theoretical models described in the Ref. 22.

BLS spectra present several distinct peaks whose frequency changes with the angle of incidence, i.e. with k_y. In fact, the detected modes have a quantized transverse wave vector k_x due to the lateral confinement in the transverse direction (x) which determines the discretization of the spectra, while the k_y value assumes continuous values determining the dispersive behaviour of the modes. In the case of stripes magnetized along their length, the frequency measured for **H** = 1.0 kOe, as shown in

Fig. 3(b), decreases on increasing k_y, i.e. on increasing the incidence angle of light, as expected for this family of modes which, resembling magnetostatic backward volume modes, have a negative group velocity. We notice that when k_y is close to zero the calculated mode frequencies tends to those measured in geometry (a) and that the overall behaviour is well reproduced by calculations. It is noteworthy that the lowest-frequency peak has the largest BLS intensity, regardless of the angle of incidence.[22]

Measurements performed with transverse stripe magnetization (**H** along the x axis, as shown in Fig. 2(c)) have been recorded by applying a static field of 2.9 kOe in order to ensure that most of the stripe volume was magnetized to saturation. BLS spectra revealed the presence of two distinct families of peaks, as seen in Fig. 3(c). One set of peaks exists at frequencies above 10 GHz and forms a dense band where the peaks are only partially resolved. Peaks of the second type exist at lower frequency, where at least two well-separated peaks are detected. First we note that the group velocity of these modes is positive, i.e. the frequencies of all peaks grow when k_y is increased. This can be understood as follows. Suppose that the magnetization was completely uniform, and define an in-plane wave vector $k_{in} = k_x \boldsymbol{e_x} + k_y \boldsymbol{e_y}$, where $\boldsymbol{e_x}$ and $\boldsymbol{e_y}$ are the unit vectors along the $\boldsymbol{x}$ and $\boldsymbol{y}$-direction, respectively. The orientation of k_{in} rotates from parallel to perpendicular to the applied field H as k_y is increased. This rotation increases the mode frequency in the same way the Damon-Eshbach surface mode frequency increases with change in propagation direction due to magnetostatic energies.[23] Note that the magnetic internal field is inhomogeneous and this results in the appearance of new low-frequency modes localized at the stripe edges. In this case the internal field is obtained by subtracting to the applied external field the demagnetizing field.[24] Remarkably the frequencies of the modes with $k_y = 0$, in Fig. 3(d), coincide with the frequencies measured in Fig. 3(c).

With these new experiments we showed that the above discussed resonant, non-dispersive, modes measured in Figs. 3(a) and 3(d), represent a particular case of a more general phenomenon of spin wave guided modes propagating along the stripe, and correspond to guided modes with vanishing longitudinal wave number k_y.

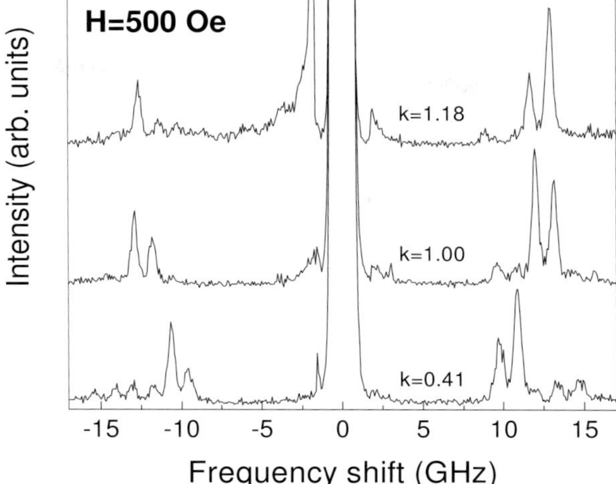

Fig. 4. Sequence of BLS spectra measured for different values of the transferred wave vector for H = 500 Oe.

4. Collective Spin Modes in Alternating Width Nano-Stripes Array

For the specimens studied in the previous paragraph the distance between the stripes of the array was relatively large (d = 500 nm), so they have been considered as isolated non-interacting objects.[1] Here we review recent experimental and theoretical investigations performed on arrays of closely-spaced magnetic stripes, where dipole-dipole interaction between adjacent stripes was important, and collective magnetostatic spin waves of the whole array, with continuous frequency spectrum, were found in a certain range of frequencies.[25-26] This is an example of propagation of collective modes in a so called "magnonic crystal" in analogy to the well established field of photonic crystals in optics and represent an emerging research field, because of the possibility of controlling the propagation of microwave spin waves.[27]

The studied specimen was an array of single layer polycrystalline Permalloy ($Ni_{80}Fe_{20}$) nano-stripes with thickness L = 30 nm and widths w_1 = 350 nm and w_2 = 500 nm (differential width $\Delta w = w_2 - w_1$ = 150 nm and array period T = 960 nm) alternated in an array. The edge-to-edge spacing between the stripes was d = 55 nm, a value sufficiently low to allow dipolar interaction between adjacent stripes. A large patterned area (4×4 mm^2) was fabricated on silicon substrate using deep ultraviolet lithography with 248 nm exposure wavelength followed by a lift-off technique. BLS experiments were performed in the scattering geometry of Fig. 2(a) where the stripes are longitudinally magnetized by an applied field **H** = 500 Oe. Figure 4 presents a series of spectra measured for different values of the incidence angle θ, i.e. different values of the in-plane wave vector k_x. Several well resolved peaks are present in the spectra but, contrarily to what previously found for isolated stripes, the peak frequency is not constant.

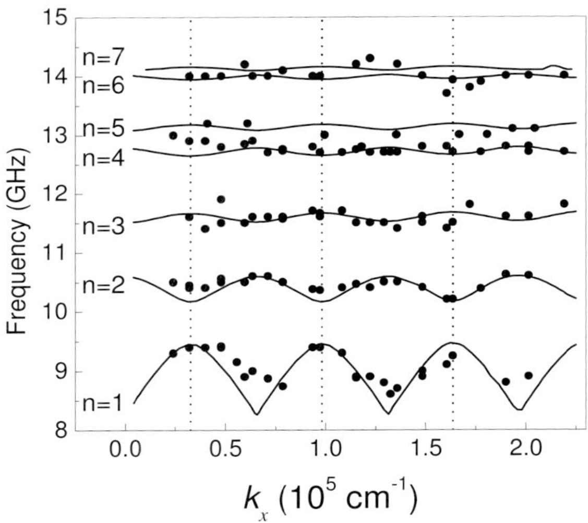

Fig. 5. Experimental and calculated spin-wave frequency dispersion for longitudinally magnetized stripes. Vertical dotted lines indicate the edges of the Brillouin zone.

This can be better seen in the synthetic graph of the measured frequencies as a function of the transferred wave vector, Fig. 5, which reveals an oscillating behaviour of the different modes, in very good agreement with the theoretical calculations. These are collective modes propagating through the array which are characterized by a periodic dispersion due to the appearance of Brillouin zones determined by the artificial periodicity. Note that the period of oscillation in k-space is inversely proportional to the periodicity in real space, as can be seen in

Ref. 26 where results for homogeneous stripes array are also presented and discussed.

As seen in Fig. 5, the oscillation amplitude is larger for the lowest frequency mode. The physical reason of this behaviour can be understood looking at the distribution of dynamic magnetization across the stripes, shown in Fig. 6. In particular, one observes that the modes, counted by the integer number *n*, differ by the number of nodes across the stripes. The amplitude is larger in narrower stripes for even modes while it is the opposite for odd modes. Calculations show that in stripes of a particular type the amplitude is larger, when the eigen-frequency is closer to one of the absolutely separated stripes of this type. This means that in this particular mode the stripes of this type are excited in resonant way, whereas the motion of magnetization of the other type is forced by

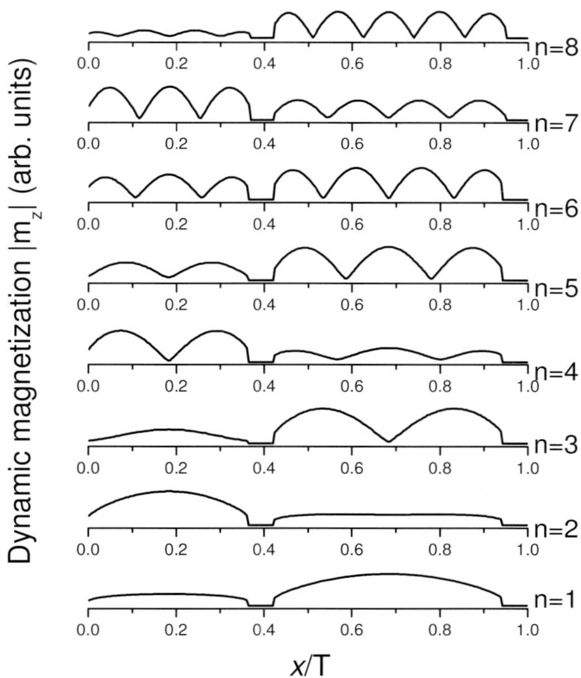

Fig. 6. Calculated distributions of dynamic magnetization for the lowest frequency modes at $k_x = 0.25 \cdot 10^5 \mathrm{cm}^{-1}$. The dimensionless coordinate in abscissa is normalized to the array period *T*. The integer number *n* labels the different modes.

the dynamic stray fields from the stripes of the former type. With this information about the dynamical magnetization distribution in hands, one can easily understand that the modes become less dispersive with increasing mode number, because the dynamic stray fields of the modes presenting several nodes drastically decreases. In fact, it is evident that the number of nodes increases with the mode number. The decrease of the stray field leads to a weaker dipole coupling of adjacent stripes and this results in a smaller dipole energy difference between the in-phase ($k_x = 0$) and the anti-phase ($k_x = \pi/\mathrm{T}$) precession of the magnetization vector.

If one looks in more details to the comparison between experiments and calculation, it can be seen that the measured oscillation amplitude is systematically smaller than the calculated one. A possible reason for this is that in the calculation the magnetic damping of the collective excitations on the array, is not taken into account, leading to an overestimation of the number of neighbouring stripes which are coupled by the dipolar interaction. The decrease of the number of coupled neighbours may result in the decrease of the frequency dispersion. In addition, as the group velocity (first derivative of the curves in Fig. 5) decreases with the mode number, the correlation length and the number of coupled stripes also decrease.

Another important observation is that the correlation length should vary significantly across Brillouin zones, because the group velocity changes from zero (at one zone edge) to maximum (at the zone centre) and back to zero at another edge. A smaller group velocity results in a larger number of magnon states in a unit frequency range. Therefore BLS intensity should increase with approaching the incidence angle corresponding to edges of Brillouin zones $k_x = (\pi \pm 2\pi n)/\mathrm{T}$.

Then, if a BLS line has a linewidth comparable with the frequency width of Brillouin zones, the dependence of BLS intensity on the magnon density of states will result in a shift of the maximum of the intensity peak towards the frequency at $k_x = (\pi \pm 2\pi n)/\mathrm{T}$. This is visible in Fig. 5: experimentally measured frequencies are slightly shifted towards the edges of Brilloin zones for all collective modes.

To complete our analysis and to verify the confidence of the theoretical model[22] the frequency dependence on the external field H strength has been also investigated. These measurements have been performed for a fixed incidence angle of 20° ($k_x = 0.81 \cdot 10^5 \mathrm{cm}^{-1}$),

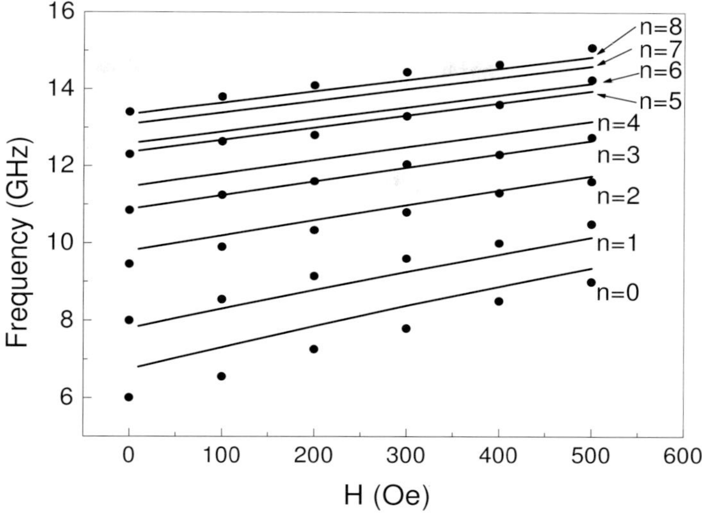

Fig. 7. Spin-wave frequencies versus applied field **H**. The mode labels are the same used in Fig. 6.

starting from **H** = 500 Oe and decreasing the field down to zero so that the stripes are always in the saturated state. The comparison between the experimental points (filled circles) and the calculated frequencies (dotted line) is depicted in Fig. 7. Remarkably, the frequency of the whole set of detected collective modes is linear in H and agreement with theoretical predictions in the whole range of magnetic fields investigated.

5. Micromagnetic Calculation of Magnetic Normal Modes in Patterned Structures

Before presenting the experimental results obtained about the effect of layering on the spin excitations of laterally confined structures, we illustrate the micromagnetic method we use to calculate the frequency and spatial distribution of modes in layered nano-elements. The three-dimensional object-oriented micromagnetic framework OOMMF micromagnetic package (version 1.2.0.4)[28] was used to simulate the dynamical properties of both of single-layer and multi-layer nano-elements which are discretized into cells of lateral dimensions of 5×5 nm^2. In practical cases, the thickness of the layers does not exceed few nanometers therefore only negligible variations of the magnetization perpendicular to the element plane can be assumed. Since we report results on permalloy ($Ni_{80}Fe_{20}$) nano-elements, magnetocrystalline anisotropy can be neglected and usually for large inter-elements separations they are considered as non-interacting. As far as the dynamical calculations are concerned, here we briefly recall the procedure for calculating the eigen-mode frequencies and profiles, while a complete description is found in Refs. 9, 29 and 30. For each applied field, the equilibrium magnetization distribution is used as starting magnetic ground state and the system is excited by an out-of-plane gaussian field pulse with peak amplitude of 10 Oe and full width at half maximum of 2 ps. To excite all the magnetic eigen-modes, gaussian field pulses with different spatial symmetries must be used. However, to avoid a large number of simulations, a pulse characterized by a "suitable" spatial asymmetry can be used. For instance, the field pulse can be built from the superposition of a uniform pulse and a pulse with either even or odd symmetry with respect to the X- and Y-directions. In addition, to excite acoustic (optic) magnetic modes, magnetic field pulses which are in phase (out-of-phase) in the two layers were used.

The time evolution of the magnetization under the Landau-Lifshitz-Gilbert (LLG) equation is recorded for each computational cell at regular time steps (Δt = 5 ps) and with a damping factor set to α = 0.01. Sometimes a smaller damping parameter (α = 0.0001) can be used to observe the magnetization oscillations over several periods because in the OOMMF code calculations are performed at T = 0 K, where thermally activated spin waves are not excited. Such a small damping parameter has, however, the disadvantage that the 2D spatial distribution which describe the profile of magnetic modes have a strong high wave vector noise superimposed in all the simulations. Note that the eigen-frequencies do not depend significantly on the damping. To reduce the calculation time we used a fifth-order Runge-Kutta integrator (rkf54s) with an error rate parameter of 1×10^{-7} which is faster and more accurate than the Euler evolver. The overall power spectrum, with a frequency resolution of 0.2 GHz, is finally obtained by summing the local power spectra of the whole set of cells. Usually several peaks are present in the power spectrum corresponding to eigen-frequencies of the system, as will be shown in the next paragraph for rectangular or circular dots. Note that the above mentioned procedure overcomes the limits of previous studies where the eigen-modes frequencies of magnetic elements are calculated performing a Fourier transform of the time evolution of the average magnetization. In this case, the modes whose average spatial dynamic magnetization is zero do not appear in the calculated power spectrum. However, these mode can be active in the BLS scattering process and give a contribution in the measured spectra because different portions of the elements are differently weighed by the phase term of the incoming light.[31] The 2D spatial distribution of each eigen-mode is achieved by plotting either the real or the imaginary part of the Fourier coefficients for each eigen-frequency. Note that both the real and the imaginary parts of the coefficients describe the same mode and have the same symmetry. However, the relative amplitude of the two components is different for the various modes, so that it is convenient to plot for each mode the largest component. To this respect, it interesting to observe that those modes which represent a resonance of the whole magnetic element have the imaginary part prevailing, in agreement with the well known fact that the susceptibility of an oscillating system would be imaginary at resonance. Please note that our approach gives similar results to the method reported in Refs. 20 and 21, where the spatial distribution of the modes is obtained by plotting the square of the amplitude of the Fourier coefficients, but has the advantage that information about the nodal planes of the modes is not missed. In reporting the mode profiles we plot the component of the magnetization vector which is perpendicular to the sample plane (m_z), because it gives the main contribution to the BLS cross-section.[31] According to the previous literature, the classification of the magnetic eigen-modes in the case of fully saturated state, where static magnetization is almost constant and parallel to the external bias magnetic field, is similar to the case continuous magnetic film, respectively.[32–34] where: (i) EM denotes the edge-modes, i.e. the modes that are localized near the particle edges in the direction of the applied magnetic field, (ii) *m*-BA denotes the dipolar modes with nodal lines perpendicular to H (backward-like modes), (iii) *n*-DE stands for modes with nodal lines parallel to the direction of H (Damon-Eshbach-like modes). The “mixed” modes (iv) are labelled *m*-BA × *n*-DE. Please note that this scheme is a useful but simplified approximation, because interlayer dipolar coupling between ferromagnetic layers in a multilayer stack can introduce a distortion of the mode profiles as well as specific features which prevent an unambiguous fit to the above classification. In addition, similarly to the acoustic and optical modes assignment in a continuous (unpatterned) trilayer consisting of two ferromagnetic films separated by a non-magnetic spacer, the calculated modes can be primarily classified according to whether the precessional motion of the dynamic magnetizations in the two ferromagnetic disks is in-phase (IP) or out-of-phase (OP).

6. Effect of Patterning on the Eigen-Mode Spectrum of Layered Structures

6.1. *Continuous film and trilayer*

To illustrate the effect of patterning (lateral confinement) on the eigen-mode spectrum of layered elements, let us have a look to the difference between the spectrum of a magnetic single layer and that of a magnetic bilayer. In Fig. 8, the BLS spectra relative to a 10 nm thick permalloy film and to a trilayer consisting of two permalloy films with a Cu interlayer, all 10 nm thick, are shown. The spectra have been measured for an applied field H = 1.0 kOe and an incidence angle of light $\theta = 10°$. For the single layer film, only one peak is present in the spectrum, corresponding to the Damon-Eshbach mode[23] while, for the unpatterned NiFe/Cu/NiFe trilayer, two peaks are observed, which correspond the acoustic and optic mode depending on wheater the precessional motion of the dynamic magnetization in the two layers is in phase or out of phase, respectively.[13] For a Cu thickness tending to infinity these two modes are degenerate and their frequency coincides with that measured for the single layer film. As a consequence of dipolar coupling, the degeneracy is removed and they assume different frequency values. For these unpatterned samples all the modes are dispersive, as can be inferred from the upper panel of Fig. 9 where the frequency values are plotted as a function of the transferred wave vector.

6.2. *Effect of lateral confinement in one in-plane direction: stripes*

To put in evidence the influence of patterning on both the acoustic and the optic spin wave modes, let us now show the results of trilayered stripes with the external field applied along their long side (interaction geometry of Fig. 2(a)). The array of stripes (width w = 470 nm and separation d = 260 nm) is fabricated from the symmetric NiFe(10 nm)/Cu(10 nm)/NiFe(10 nm) trilayer. Since the copper spacer is relatively thick any inter-layer exchange coupling between the NiFe layers is excluded.

In the case of stripes, since they are longitudinally magnetized, several dispersionless peaks are observed in the spectra (lower spectrum in Fig. 8). The discretization of the peaks is connected with the presence of resonances across the stripes and is clearly observable for the acoustic mode while it is less pronounced for the optical mode. This latter mode, in fact, is less dispersive and the frequency gap between resonant modes is more difficult to be resolved. It is interesting to note that measured frequencies are in the range where excitations of the continuous trilayer, used as reference sample, are observed, as shown in the lower panel of Fig. 9. Here we have plotted, the calculated frequencies for both the acoustic (continuous lines) and the optic mode (dashed lines) even if for the latter the measured frequencies are almost equal to that of the continuous (unpatterned) trilayer (dotted curves).

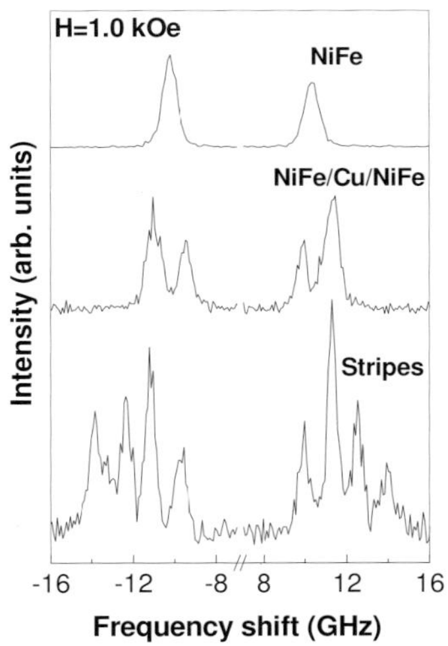

Fig. 8. Brillouin light scattering spectra measured for H = 1.0 kOe for continuous NiFe and NiFe/Cu/NiFe films and NiFe/Cu/NiFe longitudinally magnetized stripes. The scattering geometry is that of Fig. 2(a).

In Fig. 10 we present the calculated distributions of dynamic magnetization through the width of magnetic stripes for the lowest symmetric and asymmetric resonances. It is noteworthy that the shape of the modal distribution is not exactly the same: while for antisymmetric modes it very close to the sinusoidal shape, it is not so for symmetrical modes. In the latter case the curves are of more rounded shape and the effective width, at the 0.5 level, of the symmetric fundamental mode is greater than that of the asymmetric one. Another striking feature of the shown distributions is a visible difference in the effective pinning conditions for different resonances. One sees that pinning for the asymmetric mode is more pronounced, whereas for the both types of resonances it diminishes with the increase of the resonance number. An explanation of this effect is given elsewhere.[13]

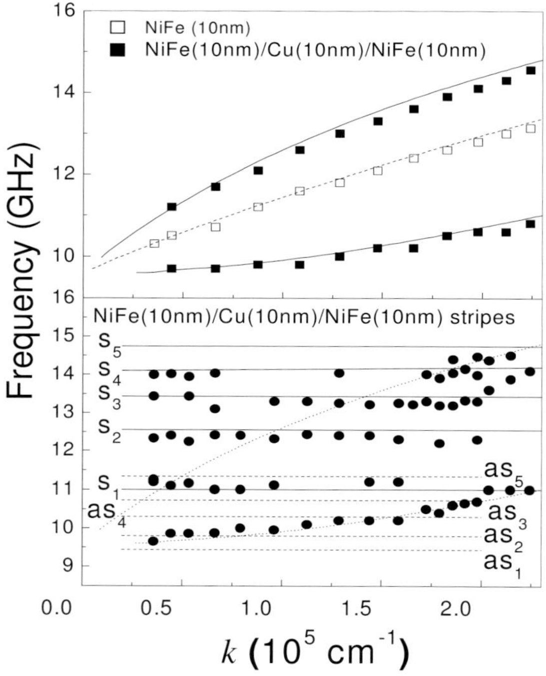

Fig. 9. Spin-wave frequencies vs transferred in-plane wave vector. Upper panel: open and full squares are the measured frequencies for continuous NiFe (10 nm thick) film and NiFe(10 nm)/Cu(10 nm)/NiFe(10 nm) continuous trilayer, respectively. Continuous and dotted curves are the calculated frequencies of the two systems. Lower panel: Dispersion for the longitudinally magnetized NiFe(10 nm)/Cu(10 nm)/NiFe(10 nm) stripes; full point are the experimental frequencies while lines are calculations. Labels (s) and (as) are for symmetric (in-phase) and asymmetric (out-of-phase) modes. Dotted curves represent the calculated frequencies of the continuous trilayer.

A deeper understanding of what happens in the trilayered stripes can be achieved extending the analysis to the case of stripes where the two magnetic layers have different thicknesses, as in the case of NiFe(30 nm)/Cu(10 nm)/NiFe(10 nm) structure. These stripes, having width w = 700 nm and edge-to-edge spacing d = 400 nm, were arranged into array of 800 × 800 μm^2. Brillouin light scattering study of thermally excited spin waves has been performed for longitudinally and transversely magnetized stripes (scattering geometries (a) and (d) of

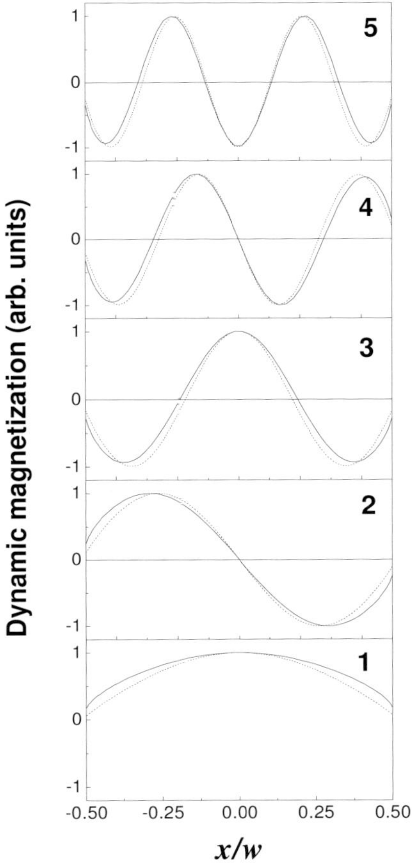

Fig. 10. Modes distribution across the stripes width for the lowest five modes. Solid and dotted curves are for symmetric (in-phase or acoustic) and asymmetric (out-of-phase or optical) precession of the dynamic magnetizations.

Fig. 2) as a function of the transferred wave vector for a fixed applied field.

Contrarily to the case of symmetric NiFe(10 nm)/Cu(10 nm)/NiFe (10 nm) stripes, the quantization of the optic mode is more evident, as shown in Fig. 11, upper panel. Note that also in this case all the modes are dispersionless, i.e. their frequency does not change as a function of the wavevector, and because of the thicker NiFe layer (30 nm instead than 10 nm) the first perpendicular standing spin waves (PSSW) appears at about 18 GHz in the spectra. Remarkably, the frequency of this mode is unaffected by patterning being equal to that measured in the reference trilayer (open circles).[13]

A complete different dispersion has been measured in the case of transversely magnetized stripes for applied field H = 2.0 kOe, as shown in Fig. 11, lower panel.

Two families of non-dispersive peaks can be clearly identified: those at higher frequency correspond to PSSW, which are resonances through the thickness of the layers, and those at lower frequency, between 6 and 13 GHz, having a magnetostatic character. For both families of modes, a shift to lower frequency values is observed, caused by the demagnetizing field which reduced the internal field (sum of the Zeeman and the demagnetizing field). A splitting of the PSSW mode with respect to unpatterned trilayers (at about 26 GHz) is also detected and is interpreted as due to the presence of two different values of the internal field existing inside the stripes.[35]

To complete our analysis the frequency dependence on the external field H has been investigated. Figure 12 presents a sequence of BLS spectra measured in the case of longitudinally (left panel of Fig. 13) and transversely (right panel) magnetized stripes for different values of the external applied field. The main features of the spectra are quite different because in the former case we observed several resolved peaks whose relative position remains almost unchanged as a function of the field intensity, while for the latter the modes merge giving rise to the formation of broad structures.

In the case of longitudinally magnetized stripes, the measurements have been performed starting from 1.0 kOe and decreasing the field down to 0 kOe so that the stripes are always in the saturated state. Remarkably, the frequency of the all modes is linear in H while for the asymmetric ones there is small curvature in the low field range. As a

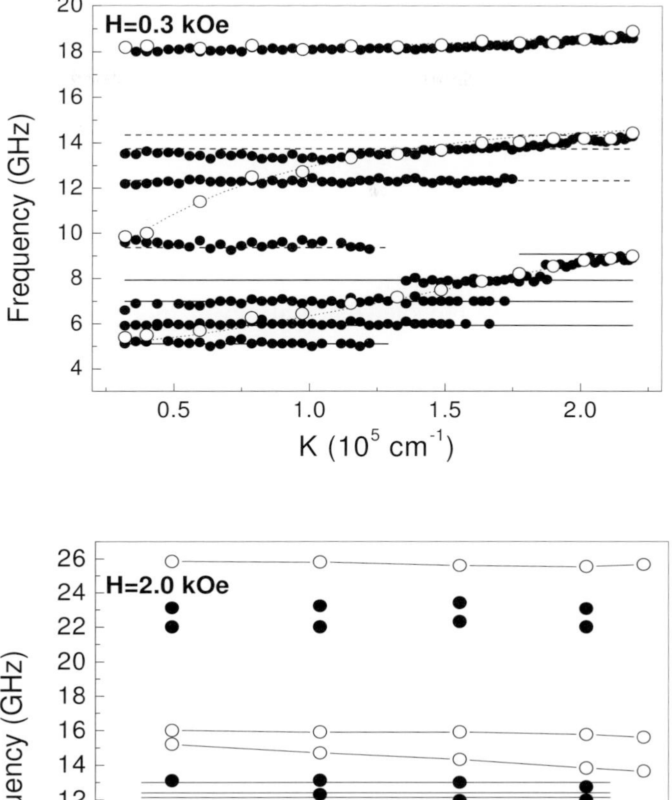

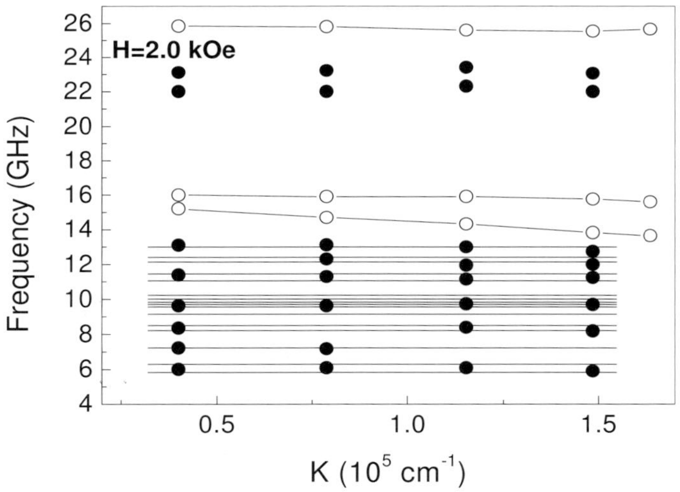

Fig. 11. Experimental spin-wave frequencies for the asymmetric NiFe/Cu/NiFe trilayered stripes as a function of the transferred in-plane wave vector k_x, perpendicular to the stripes. In the upper panel the stripes are longitudinally magnetized (H = 0.3 kOe) and solid lines correspond to optical modes and dashed lines to acoustical ones. Lower panel present the measured dispersion for transversely magnetized stripes (H = 2.0 kOe). In both panels, the measured (open points) and calculated (dotted curves) frequencies of the unpatterned symmetric trilayer are also shown for comparison.

general comment on the overall behaviour, one can note that the agreement between the measured and the calculated frequencies is rather good in the whole range of magnetic fields investigated. The comparison between the measured and the calculated frequencies as a function of the transverse magnetic field intensity is shown in right panel of Fig. 13. The measured frequencies are qualitatively reproduced by the calculation, even if a clear assignment of the mode to each measured frequency is made impossible by the large number of existing modes. In general, one can see that the number of measured modes is lower than that of the calculated one. This is due to the limited experimental resolution (0.4 GHz) as well as to the fact that only even modes are expected to give a sizable contribution in the measured spectra, as discussed in the previous paragraph. A remarkable feature which is exhibited by both the calculation and the experiment is the frequency decrease of the modes, with a minimum reached at about 800 Oe followed by an increase at smaller applied field values. This field corresponds to the value where the longitudinal hard-axis Kerr hysteresis loop is saturated.[14] Interestingly, the splitting of the perpendicular standing mode is not visible below this field, in agreement with previous investigation on transversely magnetized stripes.[35]

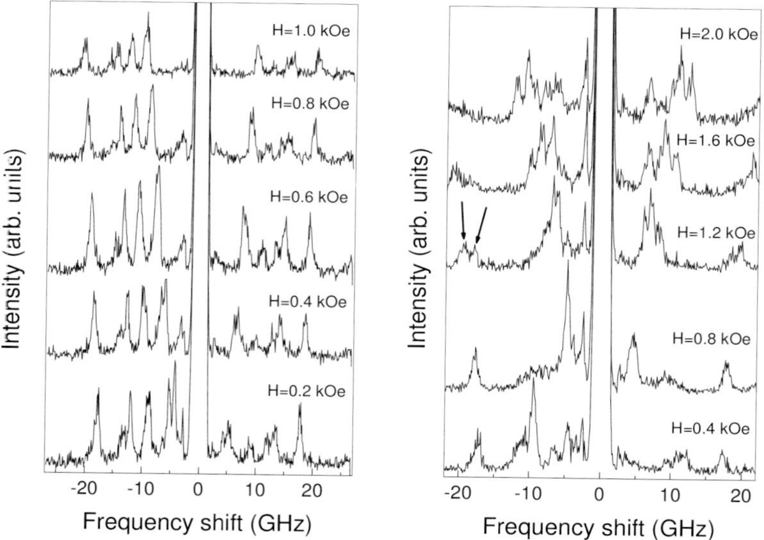

Fig. 12. Sequence of BLS spectra recorded for different values of the applied field H in the scattering configuration of Fig. 2 (a) (left panel) and (d) (right panel).

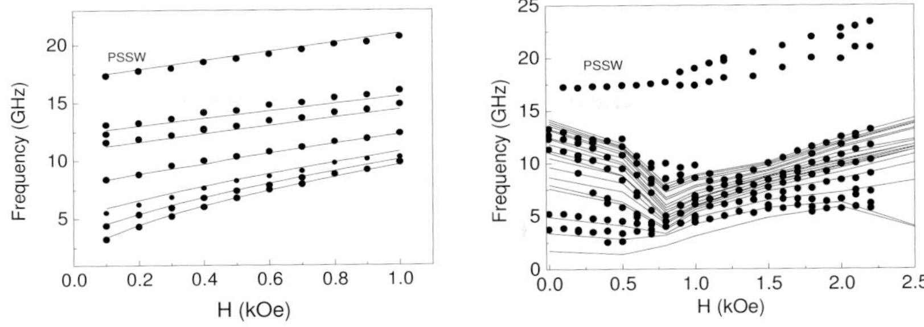

Fig. 13. Field dependence of the measured spin wave frequencies. Data refer to the spectra shown in Fig. 12.

6.3. *Effect of lateral confinement in two in-plane directions: rectangular dots*

It has been shown till now that the effect of lateral confinement is the appearance of discretized and localized modes. These phenomena can become even more evident in magnetic elements obtained by introducing a lateral confinement along both in-plane directions, for example in the case of rectangular magnetic dots. Here we show the results relative to 720 × 430 nm^2 rectangles, with a separation d = 260 nm, fabricated from the previously described symmetric NiFe(10 nm)/Cu(10 nm)/NiFe(10 nm) trilayer.

As shown in Fig. 14, the measured spectrum for H applied parallel to the long side of the rectangular dots (and the *k* perpendicular to H, as in Fig. 2(a)) is different from that measured in the case of stripes. The presence of two finite in-plane dimensions causes the presence of more peaks than those observed for trilayered stripes. In addition, the peaks are at lower frequency because the demagnetizing field reduces the internal field felt by precessing spins.

In Fig. 15, we show the measured and the calculated frequencies. The main effect of the 2D confinement is the overall reduction with respect to the corresponding stripes (see Fig. 9) of the measured frequencies, with the appearance of modes at frequency lower than 6 GHz. Similarly to the case of the stripes, also in the case of rectangles the measured spin modes are stationary waves, but their number is larger because of the coexistence of modes having nodal planes both parallel (BA-modes)

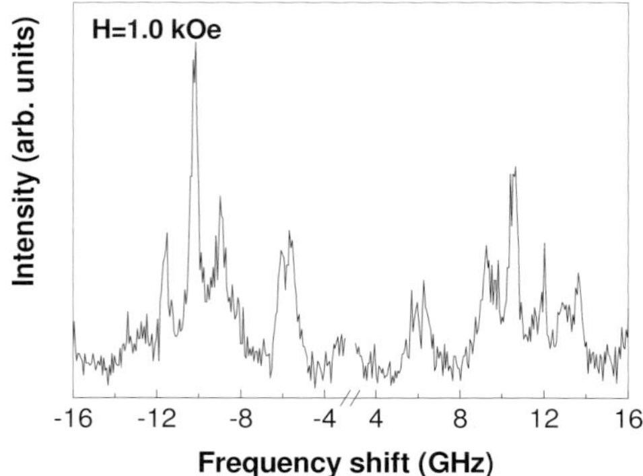

Fig. 14. BLS spectrum relative to an array of rectangular dots. The field applied is 1.0 kOe oriented along the dot's long side of the rectangular dots and the incoming wave vector of light is oriented along the short side of the element (scattering geometry of Fig. 2(a)).

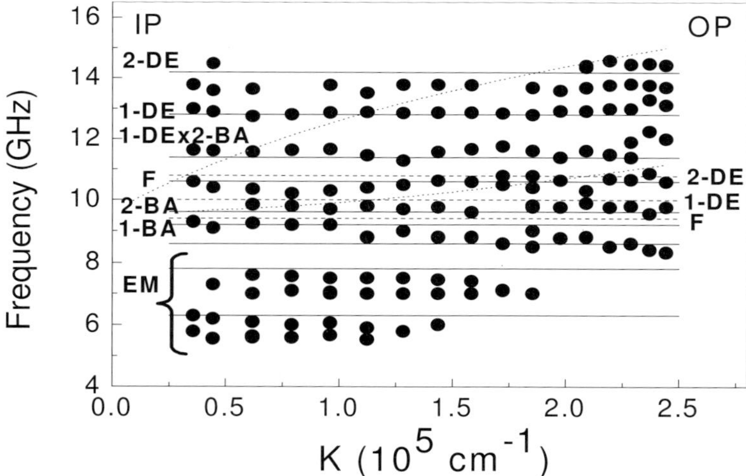

Fig. 15. Frequencies of spin-wave modes of an array of NiFe/Cu/NiFe rectangular elements (720×430 nm^2) in an applied field of H = 1.0 kOe directed along the long side of the elements measured by BLS as a function of the transferred wave vector. Dotted curves are the calculated frequencies for the NiFe(10 nm)/Cu(10 nm)/NiFe(10 nm) continuous trilayer while the horizontal lines are the calculated frequencies for IP (continuous lines) and OP (dotted curves) modes. Left (right) labels refer to the IP (OP) modes.

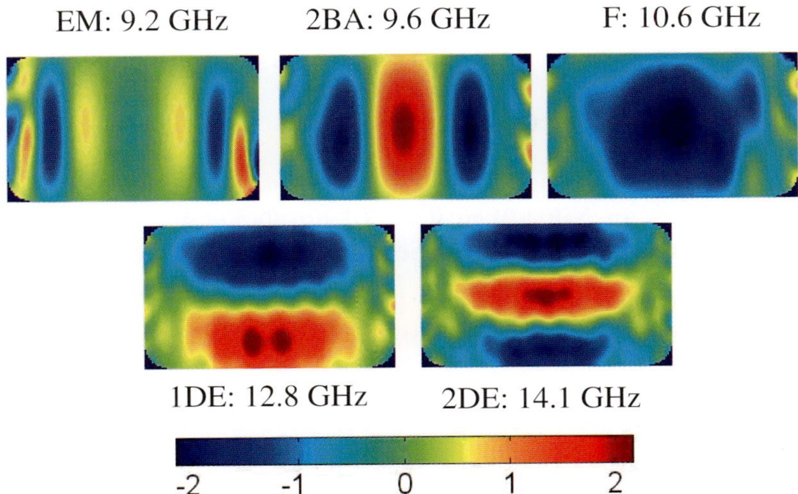

Fig. 16. Snapshots of the dynamical out-of-plane component of the magnetization m_z for some modes. The bias field H is applied parallel to the long side of the rectangular elements.

and perpendicular (DE-modes) to the direction of the applied field, as can be inferred from the analysis of our micromagnetic calculation of modes frequency and spatial profiles. Note that most of the observed peaks can be interpreted as due to IP modes (which are expected to have a larger cross section, as seen also for the continuous trilayer) while the most efficient OP modes are packed in a narrow frequency range around 10.0 GHz for rectangles. Concerning the low frequency modes, we notice that they belong to a family of localized modes existing close to the rectangular dots edges where the internal field is inhomogeneous and differ by the number of nodal plane in the direction perpendicular to that of the external applied field (short axis of the rectangular dot).

The spatial distribution of the spin wave modes calculated using the micromagnetic approach are shown in Fig. 16 for the modes with higher BLS cross-section At higher frequency a couple of backward-like modes are present below the fundamental mode whose calculated frequency (10.6 GHz) is in good agreement with the frequency position of the most intense peak (10.3 GHz) observed in the spectrum shown in Fig. 14. The measured peaks above the fundamental modes correspond in the calculation to hybridized and pure DE modes. It is important to observe that both for the F and the DE modes, the magnetization oscillation occurs in the central part of the element where the internal field is homogeneous while in the remaining part of the element it is almost null. To improve the agreement between experiment and calculation, we have used as input file for dynamical simulation the dot geometry taken from the SEM image which reveals slightly rounded corners of the rectangular dots.

7. Spin Waves in NiFe(30 nm)/Cu(10 nm)/NiFe(30 nm) Dots with Circular and Elliptical Cross Section

In this final paragraph we present the results of our BLS study of the dynamical properties of NiFe(10 nm)/Cu(10 nm)/NiFe (10 nm) dots with elliptical cross-section and different eccentricity. Our aim is to systematically study the effect of the in-plane shape anisotropy on the high frequency magnetization dynamics. The variation of the eccentricity ε is obtained by changing the major axis *a* of the ellipse, which assumes the values 200, 300, 400, 500, and 600 nm, and leaving the minor axis *b* fixed to 200 nm, so that the dot eccentricity ε = *a*/*b was* $\varepsilon = 1$, 1.5, 2, 2.5 and 3. Details about sample preparation are described elsewhere.[9]

7.1. *Circular dots*

We start with the presentation of experimental results obtained for circular dots ($\varepsilon = 1$). The hysteresis curve measure by MOKE (Fig. 17) is characterized by an almost anhysteretic behaviour which can be well reproduced by our micromagnetic calculation. The sample is initially saturated in a positive field H = 1.5 kOe, corresponding to a parallel alignment of the layers magnetizations. These remain aligned as the field intensity is decreased down to 0.6 kOe, but when the field is further reduced, the magnetizations of the two layers start to rotate in opposite directions until they reach a perfect anti-parallel alignment at zero field. This antiparallel configuration at remanence is clearly due to the dipolar interaction between the layers, which forces them to be antiparallel in the absence of field. It is important to notice that, during the reversal, the magnetization in each layer remains almost in single-domain state. This is different from the recent findings of other works on thicker trilayers,[36] where evidence of vortex formation in the disks was found.

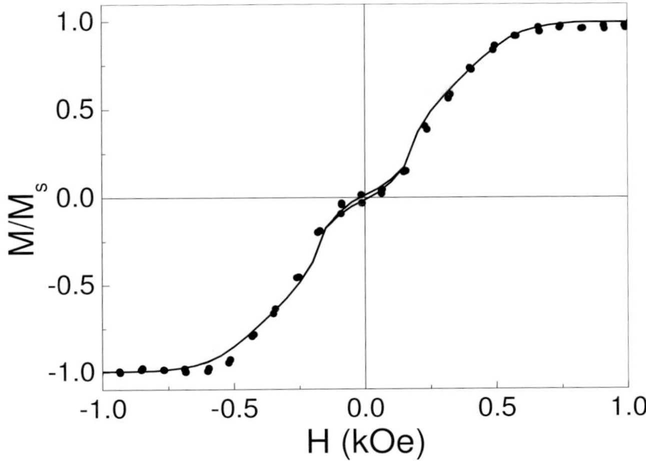

Fig. 17. Measured longitudinal MOKE (full points) and simulated hysteresis loop (continuous curve) for the trilayered circular dots.

To study the field evolution of modes profiles, we have measured BLS spectra for different values of the applied magnetic field ,in the range 0–1.5 kOe, and k perpendicular to H, as in Fig. 2(a). On both the Stokes and the anti-Stokes sides of the spectrum several well resolved peaks are seen in Fig. 18, whose relative intensity changes as a function of the external applied field. On reducing the field intensity, the frequency of all the modes monotonously decreases and reaches a minimum at about 0.6 kOe, which corresponds to the saturation field (see Fig. 17). On further decreasing the field intensity, the frequency of the modes starts to increase again, reflecting the simultaneous and opposite rotation of the magnetizations of the two layers. The observed modes are dispersionless, i.e. their frequency does not change as a function of the incidence angle of light, as typical for resonant modes in a laterally confined magnetic system. Moreover, no angular dependence of the spin mode frequency on the in-plane direction of the applied magnetic field has been observed for saturated dots. This enables one to exclude any appreciable in-plane anisotropy of the dots, as expected for isolated, circular dots of permalloy.

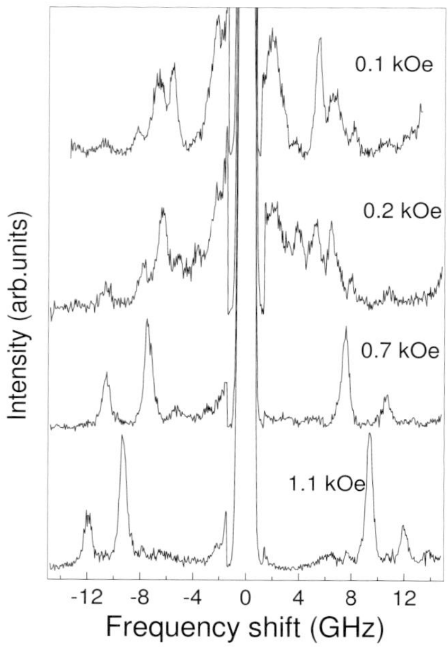

Fig. 18. Sequence of BLS spectra recorded for different values of external magnetic field.

Figure 19 reports the average power spectra for the NiFe/Cu/NiFe disk-shaped dots, calculated for an out-of-plane (Z-direction) field pulse in presence of an 1.5 kOe uniform dc magnetic field applied along the X-direction. First of all, we have verified that applying excitation pulses with different symmetries in the XY plane allows the excitation of different normal modes. For example, either the DE-like or the BA-like modes with an odd number of nodal lines, can be efficiently excited by using a perpendicular pulse with odd symmetry with respect to the Y and X axis, respectively. In the top panel of Fig. 19 the calculated profiles of the excited normal modes for the bottom and top NiFe dots are shown.

Here one can notice that the 1-DE×3-EM is excited by the Y-odd pulse and it is almost absent for the other pulse symmetries while the 3-BA×4-EM is excited by the X-odd pulse, only. These modes are not present at all or are very small in the power spectrum obtained for the system excited by the uniform pulse. Both the fundamental and the localized modes are excited by pulses with any symmetry. These latter

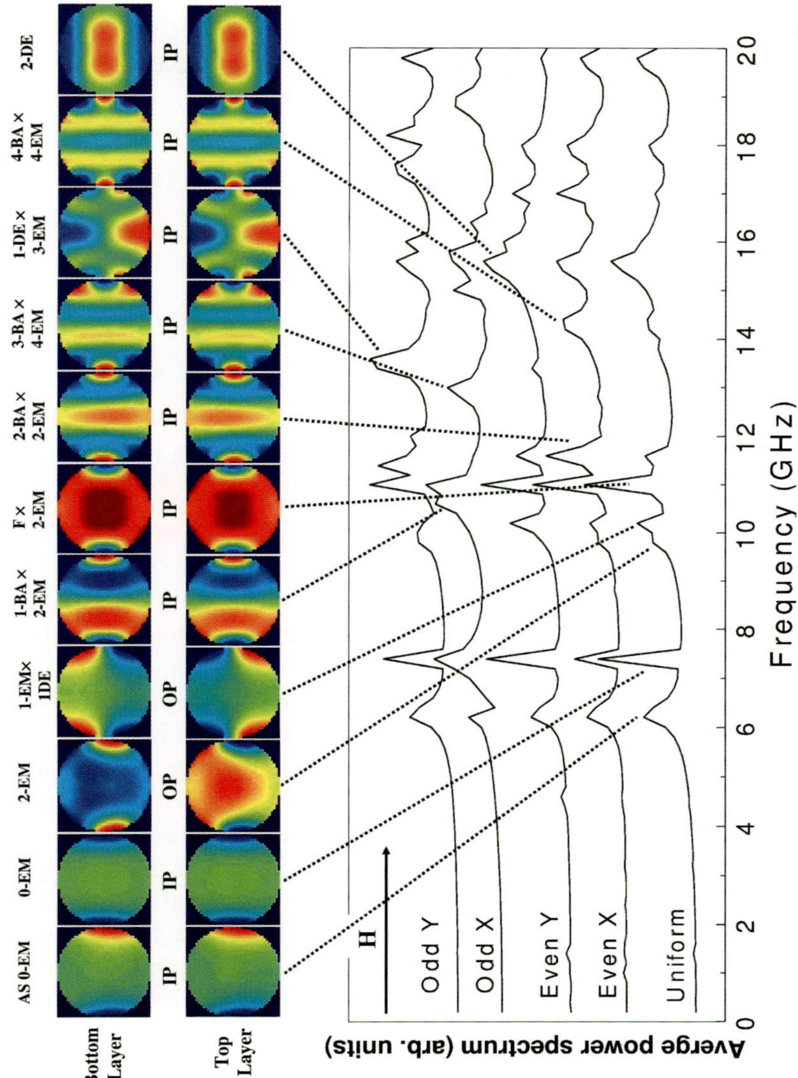

Fig. 19. Calculated average power spectra obtained with an external static field $H = 1.5$ kOe for different excitation pulses as described in the labels, i.e., for uniform pulse, and for X-odd symmetric and Y-odd symmetric pulses. The modes profiles in the two layers as well as the direction of the applied magnetic field H are shown in the upper part.

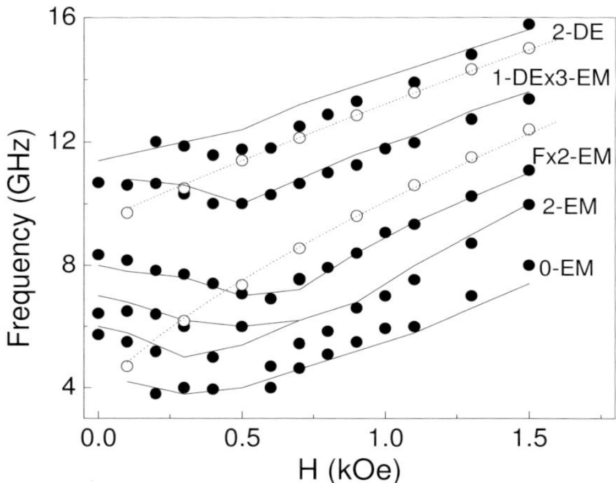

Fig. 20. Measured (full points) and calculated (solid curves) frequencies of the spin wave modes supported by the multilayered circular dots. Open points and dotted curves refer to the case of the continuous trilayer.

are edge modes (0-EM) with the difference that the former is asymmetric (AS) inside each layer (with zero net dynamic magnetization) while the latter is symmetric (non zero net dynamic magnetization).

Assuming that the main contribution to the cross-section is given by the top layer, which is quite reasonable because of the large attenuation of light reaching the bottom layer, and reminding the qualitative discussion presented in Ref. 31, we found that only a few modes, namely the 0-EM, 2-EM, F×2-EM, 1-DE×3-EM and 2-DE, are active in the scattering process and give a substantial contribution in the BLS spectra. For this reason, in the comparison with the experimental frequencies, we have followed the field dependence of the above mentioned five modes, only. For all these modes the dynamic magnetizations in the two layers oscillates in-phase with the exception for the 2-EM which is out-of-phase. Please note that the modes labelled as 1-DE and 2-DE correspond in Ref. 9 to the modes labelled as 2-DE and 3-DE, respectively.

A satisfactorily agreement is found between experiment and calculation, as shown in Fig. 20, with the exception of the modes which are localized near the edges of the magnetic elements, such as the 2-EM and the 0-EM mode. This is not surprising, since the frequencies of these modes are influenced to a large extent by the real morphology and curvature of the dot edges. This is reflected by the pronounced broadening of the experimental BLS peak of this mode observed in the spectra of Fig. 18. In addition, in micromagnetic calculations, the dot edge is discretized, so that we expect the agreement between measured and calculated frequencies to be the worst where the discretized edge fails to mimic satisfactorily the actual rounded shape of the dot.

To gain further insights into the field evolution of the modes character, we present in Fig. 21 the profiles of the dynamic magnetization in the top and bottom NiFe disks for the F×2-EM and the 0-EM modes at different values of *H*. The character of these mode is particularly easy to identify, because it remains unchanged over the whole field range investigated. It is very interesting to observe how the localization region of the mode strictly follows the magnetization configuration of the layers and reflects their in-plane rotation. A similar in-plane rotation of the mode profiles occurs also for other modes. However, as soon as the magnetization in the two layers deviates from parallel alignment, it is not possible to follow a well-defined profile vs *H*, because the character of each mode becomes less and less defined and mode-crossings and different degrees of hybridization occur.

7.2. *Elliptical dots*

In the remaining of this chapter we extend our investigation to the case of trilayered elements with elliptical cross section.[37] Typical sequences of measured Brillouin light scattering spectra are shown for different dots eccentricity ($\varepsilon = 1$, 1.5 and 2.5) in Figs. 22(a) and 22(b), for an external field ($\mathbf{H}$ = 1.5 kOe) applied along the easy and hard magnetization direction of the elements, respectively. Several discrete peaks can be recognised and their frequency positions are reported in Fig. 23. It can be noticed that, for H parallel to the easy (hard) direction, there is an overall increase (decrease) in the measured frequencies for increasing eccentricity with a saturation for $\varepsilon > 2$. The dependence of the measured spin mode frequencies as a function of the in-plane direction (ϕ) of the applied magnetic field with respect to the easy axis of the ellipses is shown in Fig. 24 for the pillars with eccentricity $\varepsilon = 1$, 1.5, and 2.5. The flat dispersion curves for $\varepsilon = 1$ reflect the expected in-plane isotropic behavior of a cylindrical pillars. For eccentricity values larger than 1, an appreciable in-plane uniaxial anisotropy appears, due to the effective

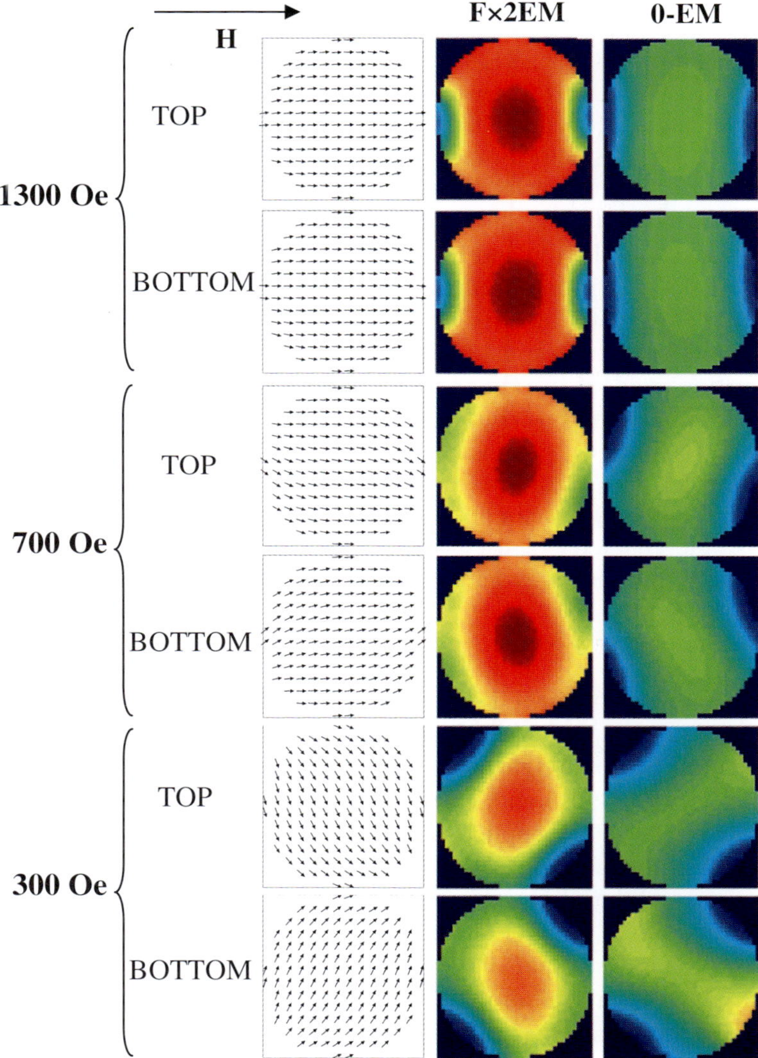

Fig. 21. Magnetization profiles of the dynamic magnetization for the F×2-EM and the 0-EM mode in the two (top and bottom) NiFe disks for different values of the external field H applied along the arrow direction. The magnetic profiles are calculated using the procedure described in the text. On the first left column the static magnetization distributions in the two layers are also shown.

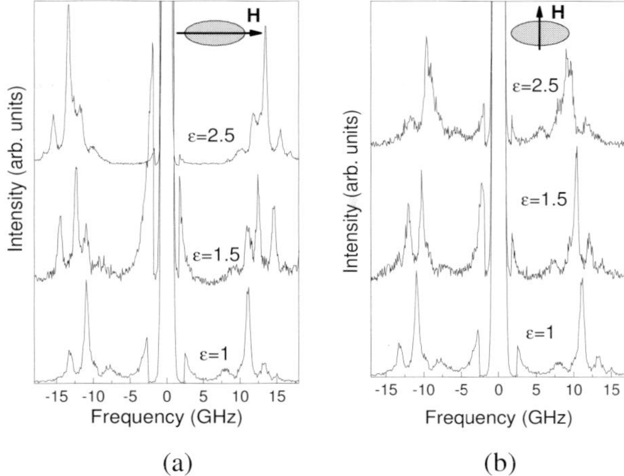

Fig. 22. Set of Brillouin spectra taken with the applied field **H** along the easy (left panel) and the hard (right panel) magnetization axes for dots with different eccentricity.

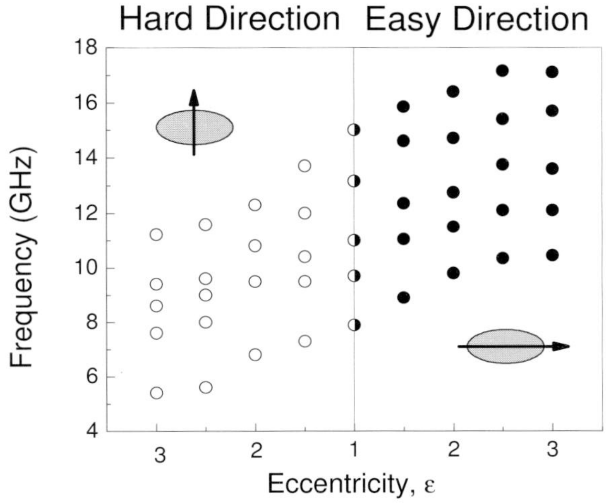

Fig. 23. Experimental Brillouin frequencies of various modes plotted versus the eccentricity. The results are shown for the applied field along the easy (full points) and hard axes (open points).

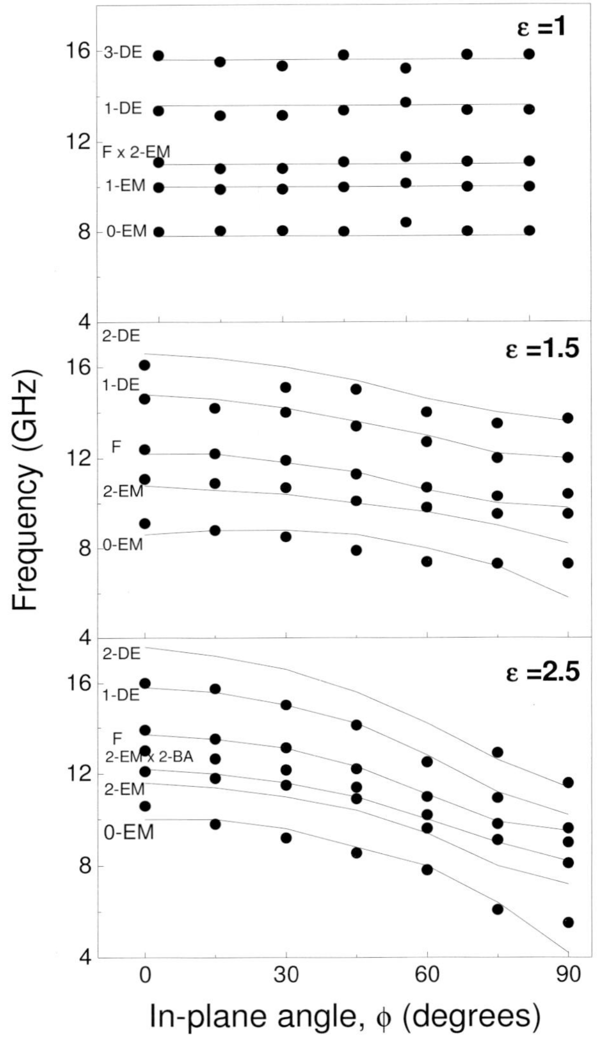

Fig. 24. Experimental Brillouin frequencies (full symbols) and calculated (solid curves) frequency values for ellipses of different eccentricity, measured versus the angle (ϕ) between the applied field and the major axis of the ellipses.

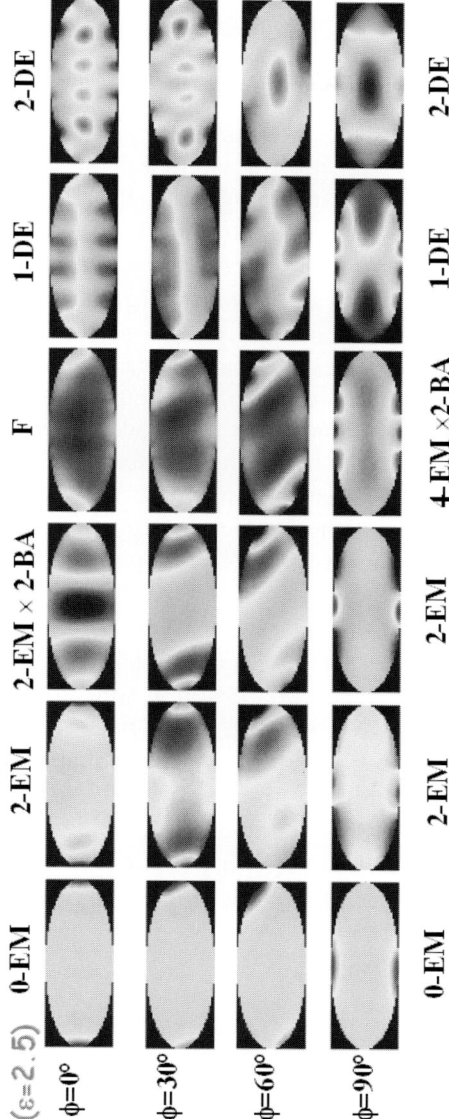

Fig. 25. Calculated out-of-plane component of the dynamic magnetization of various spin modes plotted versus the angle ϕ between the applied field and the major axis of the ellipses (eccentricity $\varepsilon = 2.5$). The labels on the top and bottom rows are assigned to the modes to achieve a tentative classification of each of them according to its symmetry.

demagnetizing factor associated to the shape of the pillars. As for the dependence on the direction of the in-plane applied field H, the demagnetizing field increases with ϕ, since one is approaching the hard axis, producing a decrease in the effective magnetic field and a consequent reduction of the mode frequencies. All these features are well reproduced by the micromagnetic calculations (solid curves in Fig. 24) with significant discrepancy appearing only for the lowest-frequency mode at $\phi = 90°$. It is interesting to briefly illustrate in detail the evolution of the mode profiles due to the variation of the in-plane direction of the magnetic field ϕ, shown in Fig. 24 for the sample with $\varepsilon = 2.5$. Note that only the profiles relative to the topmost layer are plotted, because this layer gives the main contribution to the BLS cross section; moreover, the mode profiles of the bottom layer are almost the same as that of the topmost one. First of all we notice that, the modes labels are slightly different from those used in Ref. 37 and the magnetization distributions are more smooth because we have plotted the imaginary or the real part of the Fourier coefficients depending on their relative intensity and because of a different algorithm used to plot these profiles. We observed that the 2-EM mode gives an appreciable contribution while it is not BLS-active for $\varepsilon = 1$.

Different from all the other measured modes, this mode is characterized by an out-of-phase precession of spins in the two elemental layers for values of ϕ below about 30° while it is in-phase for ϕ larger than 60°. As shown in Fig.25 the mode character changes with ϕ because mode crossing and hybridization effects occur. For this reason at intermediate values of ϕ it is not simple to unambiguously assign a label to each mode. We notice, however, that for $\phi = 0°$ and 90° (principal symmetry direction of the ellipses) the mode symmetry remains almost completely unaffected.

8. Conclusions

In this chapter, we have reviewed relevant experiments about the dynamical properties of layered magnetic nano-elements, carried out by our group at the Department of Physics of the University of Perugia by using the Brillouin light scattering technique.

We have examined experimentally and theoretically families of propagating guided modes which exist in isolated (non interacting) magnetic stripes. The modes have continuous wave numbers along the direction of propagation and are quantized in the transverse direction due to lateral confinement. When the stripes are coupled by dipolar interaction, magnetic modes have a dispersive character and present an oscillating behaviour encompassing several Brillouin zones, induced by the artificial periodicity of the array. Each mode exists in its frequency range, separated from the neighbouring by a prohibited zone.

We have then explored the effect of the lateral confinement on the spin wave properties of layered structures showing that the principal effect of the inter-layer dipolar coupling is the formation of either "acoustic" (in-phase) or "optic" (in anti-phase) coupled spin wave modes. In addition, when the system is confined in the two in-plane directions, additional mode-splitting and an overall reduction of the measured frequencies has been observed with the appearance localized modes.

Finally, we have shown that with the BLS technique it is possible to resolve several spectral features related to quantized spin mode excitations in ordered arrays of nanometric layered dots of elliptical shape. The data were recorded as function of the eccentricity of the ellipses and of the in-plane direction of an applied magnetic field. The interpretation of the data required a calculation of the magnetic normal modes and of the BLS cross section which allow us to assign each peak in the spectra to a corresponding spin mode. When the in-plane direction of the applied field varies, the magnetization shows a remarkable anisotropy, proportional to the eccentricity, which produces a strong dependence of the spin mode frequency on the field direction. Also the magnetization profile of most of the spin modes changes as a function of the direction of the applied field, and in many cases even the mode character and symmetry are affected.

References

1. S. O. Demokritov and B. Hillebrands, in *Spin Dynamics in Confined Magnetic Structures I*, Eds. B. Hillebrands and K. Ounadjela, *Topics in Applied Physics* **83** (Springer-Verlag, 2002), p. 65; and references therein.
2. G. Leaf, H. Kaper, M. Yan, V. Novosad, P. Vavassori, R. E. Camley and M. Grimsditch, *Phys. Rev. Lett.* **96**, 017201 (2006).
3. F. Montoncello, L. Giovannini, F. Nizzoli, P. Vavassori, M. Grimsditch, T. Ono, G. Gubbiotti, S. Tacchi G. and Carlotti, *Phys. Rev.* ***B*76**, 024426 (2007).
4. *Magnetic Nanostructures*, edited by H. S. Nalwa (American Scientific), Los Angeles, (2002).

5. G. Carlotti and G. Gubbiotti, *J. Phys. Condens. Matter* **14**, 8199 (2002).
6. S. O. Demokritov, *J. Phys. Condens. Matter* **15**, S2575 (2003).
7. G. Gubbiotti (Ed.), *Magnetic Properties of Laterally Confined Nanometric Structures* (Transworld Research Network, 2006).
8. L. Giovannini, F. Montoncello, R. Zivieri and F. Nizzoli, *J. Phys. Condens. Matter,* **19**, 225008 (2007).
9. G. Gubbiotti, M. Madami, S. Tacchi, G. Carlotti and T. Okuno, *Phys. Rev. B* **73**, 144430 (2006).
10. K. W. Chou, A. Puzic, H. Stoll, G. Schütz, B. Waeyenberge, T. Tyliszczak, K. Rott, G. Reiss, H. Brückl, I. Neudecker, D. Weiss and C. H. Back, *J. Appl. Phys.* **99**, 08F305 (2006).
11. D. V. Berkov and N. L. Gorn, *Appl. Phys.* **103**, 053908.
12. K. Yu Guslienko, K. S. Buchanan, S. D. Bader and V. Novosad, *Appl. Phys. Lett.* **86**, 223112 (2005).
13. G. Gubbiotti, M. Kostylev, N. Sergeeva, M. Conti, G. Carlotti, T. Ono and A. Stashkevich, *Phys. Rev.* ***B*70**, 224422 (2004).
14. G. Gubbiotti, G. Carlotti, T. Ono and Y. Roussigné, *J. Appl. Phys.* **100**, 023906 (2006).
15. N. A. Sergeeva, S. M. Chérif, A. A. Stashkevich, M. P. Kostylev, M. P. and J. Ben Youssef, *J. Magn. Magn. Mater.* **288**, 250 (2005).
16. S. I. Kiselev, J. C. Sankey, I. N. Krivorotov, N. C. Emley, R. J. Schoelkopf, R. A. Buhrman and D. C. Ralph, *Nature* **425**, 380 (2003).
17. S. I. Kiselev, J. C. Sankey, I. N. Krivorotov, N. C. Emley, M. Rinkoski, C. Perez, R. A. Buhrman and D. C. Ralph, *Phys. Rev. Lett.* **93**, 036601 (2004).
18. J. Sun, *Nature* 425, 359 (2003).
19. http://ghost.fisica.unipg.it
20. J. R. Sandercock, in *Light Scattering in Solids III*, Eds. M. Cardona and G. Güntherodt, Vol. 51 of *Topics in Applied Physics* (Springer, Berlin, 1982).
21. C. Mathieu, J. Jorzick, A. Frank, S. O. Demokritov, A. N. Slavin, B. Hillebrands, B. Bartenlian, C. Chappert, D. Decanini, F. Rousseaux and E. Cambril, *Phys. Rev. Lett.* **81**, 3968 (1998).
22. M. P. Kostylev, G. Gubbiotti, J-G. Hu, G. Carlotti, T. Ono and R. L. Stamps, *Phys. Rev. B* **75**, 054422 (2007).
23. W. Damon and J. R. Eshbach, *J. Phys. Chem. Solids* **19**, 308 (1961).
24. R. I. Joseph and E. Schlomann, *J. Appl. Phys.* **36**, 1579 (1965).
25. G. Gubbiotti, S. Tacchi, G. Carlotti, P. Vavassori, N. Singh, S. Goolaup, A. O. Adeyeye, A. Stashkevich and M. Kostylev, *Phys. Rev.* ***B*72**, 224413 (2005).
26. G. Gubbiotti, S. Tacchi, G. Carlotti, N. Singh, S. Goolaup, A. O. Adeyeye and M. Kostylev, *Appl. Phys. Lett.* **90**, 092503 (2007).
27. S. A. Nikitov, Ph. Thailades C. S. Tsai, *J. Magn. Magn. Mater.* **236**, 320 (2001).
28. http://math.nist.gov/oommf/
29. R. D. McMichael and M. D. Stiles, *J. Appl. Phys.* **97**, 10J901 (2005).

30. L. Torres, G. Finocchio, L. Lopez-Diaz, E. Martinez, M. Carpentieri, G. Consolo and B. Azzerboni, *J. Appl. Phys.* **101**, 053914 (2007).
31. G. Gubbiotti, G. Carlotti, T. Okuno, M. Grimsditch, L. Giovannini, F. Montoncello and F. Nizzoli, *Phys. Rev. B* **72**, 184419 (2005).
32. L. Giovannini, F. Montoncello, F. Nizzoli, G. Gubbiotti, G. Carlotti, T. Okuno, T. Shinjo and M. Grimsditch, *Phys. Rev. B* **70**, 172404 (2004).
33. G. Gubbiotti, M. Conti, G. Socino, G. Carlotti, P. Candeloro, E. Di Fabrizio, K. Yu Guslienko, A. Andre and A. N. Slavin, *J. Phys: Condens. Matter* **16**, 7709 (2004).
34. G. Gubbiotti, M. Madami, S. Tacchi, G. Carlotti, A. O. Adeyeye, S. Goolaup, N. Singh and A. N. Slavin, *J. Magn. Magn. Mater.* **316**, e338 (2007).
35. J. Jorzick, S. O. Demokritov, B. Hillebrands, M. Bailleul, C. Fermon, K. Yu Guslienko, A. N. Slavin, D. V. Berkov and N. L. Gorn, *Phys. Rev. Lett.* **88**, 047204 (2002).
36. K. S. Buchanan, K. Yu Guslienko, A. Doran, A. Scholl, S. D. Bader and V. Novosad, *Phys. Rev. B* **72**, 134415 (2005).
37. G. Gubbiotti, M. Madami, S. Tacchi, G. Socino, G. Carlotti and T. Ono, *J. Appl. Phys.* **101**, 09F502 (2007).

Chapter 3

NON-UNIFORM MAGNETIZATION DYNAMICS IN ULTRA-SMALL FERROMAGNETIC PLANAR ELEMENTS

V. V. Kruglyak[*], P. S. Keatley[†], R. J. Hicken[‡]

School of Physics, University of Exeter
Stocker road, Exeter, EX4 4QL, United Kingdom
[*]*E-mail: V.V.Kruglyak@exeter.ac.uk*
[†]*E-mail: P.S.Keatley@exeter.ac.uk*
[‡]*E-mail: R.J.Hicken@exeter.ac.uk*

J. R. Childress[§] and J. A. Katine[¶]

Hitachi Global Storage Technologies, San Jose Research Center
3403 Yerba Buena Road, San Jose, CA 95135, USA
[§]*E-mail: Jeff.Childress@hitachigst.com*
[¶]*E-mail: Jordan.Katine@hitachigst.com*

We present studies of the magnetization dynamics of nanoscale magnetic elements of square shape. Particular emphasis is placed upon effects that do not exist or have not been observed in microscale magnetic elements. The experiments revealed a crossover from a two mode to a single mode regime when the element size or the strength of the bias magnetic field were decreased, or when the orientation of the bias magnetic field was rotated towards the diagonal of the elements. Comparison with micromagnetic simulations allows us to understand the dependence of the mode frequencies and amplitudes upon the spatial distribution of the magnetization and effective field within the element. The simulations reveal the important role played by the exchange interaction and the edge conditions. They also demonstrate the limited applicability of the macrospin model and the concept of the wave vector in the description of high frequency precessional dynamics in nanoscale magnetic elements.

1. Introduction

There are many ways to excite precessional motion in a ferromagnet. One may use harmonic[1] or pulsed[2] magnetic fields, ultrashort optical pulses,[3] or even constant spin polarized current.[4] Independent of the nature of excitation, the resulting magnetization dynamics are described using the same universal language of magnons (spin waves) – elementary collective excitations of spins.[5] Several decades of experimental and theoretical research have demonstrated that magnons exhibit most of the properties exhibited by waves of other origins. In particular, the excitation and propagation,[1,6] reflection and transmission,[7] focusing,[8] and interference[9] of magnons were observed, as well as the Doppler effect,[10] the formation of magnonic solitons,[11] and the opening of magnonic band gaps in the spectrum of one-,[12,13] two-,[14,15] and three-dimensional[16] magnonic crystals (materials with periodically modulated magnetic properties, e.g. arrays of nanomagnets). By analogy with other quasiparticles (e.g. photons, phonons, plasmons, etc.), this area of research was called magnonics,[17] and has received increased attention each year.

Due to the recent progress in nano-fabrication, it has become possible to study spin waves in thin film magnetic elements of sub-micron dimensions.[18–21] From a technological point of view, a proper understanding of the magnetization dynamics in such elements underpins the operation of future recording head sensors, magnetic random access memory, and spin-torque devices. From a fundamental point of view, the main question to address is whether this quantitative change in dimensions can result in qualitatively new dynamic magnetic phenomena. In other words, it is necessary to establish the limits of applicability of the concepts developed in the studies of micron sized magnetic elements. Let us briefly consider the concepts that were the focus of recent magnetism research together with the reasons why they might be expected to fail in the nano-scale regime.

The first such concept is the assumption that the frequency of a spin wave is determined by its wave vector. According to Noether's theorem,[22] there is a fundamental connection between the symmetry of a physical system and the conservation laws governing its behavior. For instance, time invariance implies that energy (frequency) is conserved, while translational invariance (spatial uniformity) implies that linear momentum (wave vector) is conserved. In the case of spin waves in thin magnetic films, this means that they can be described by their in-plane wave vectors. However, the translational invariance in the direction perpendicular to the plane of a thin film is broken near its surfaces. This leads to quantization of the component of the wave vector perpendicular to the plane, i.e. the wave vector can have only a discrete number of values determined by the boundary conditions at the surfaces of the film. In a thin film element, the in-plane translational invariance is broken near the edges, and so all three components of the wave vector are quantized.[23] In addition, the demagnetizing field results in a non-uniform static effective magnetic field and magnetization everywhere within non-ellipsoidal magnetic elements of finite size. Hence, it becomes impossible to ascribe any constant wave vector to a particular mode. Moreover, a coordinate dependent wave vector, such as that introduced in Ref. 24 for a long stripe, is inappropriate in the case of true 3-dimensional confinement because it is impossible to assign a direction to the wave vector. The magnitude of the demagnetizing field increases when the aspect ratio (size to thickness) of the element is reduced. So, for the same thickness, the concept of the wave vector may still work approximately for micron sized elements but should fail for nanosized magnetic elements. However, we note that these considerations may be inapplicable when the size of the element is reduced below the exchange length, where the exchange rather than magnetostatic interaction will play the dominant role.

The assumption that the magnetostatic interaction dominates the magnetization dynamics in magnetic elements is another concept that works well in micron-sized elements but fails in nano-sized magnetic elements. Indeed, the contribution of the exchange interaction to the frequency of spin waves increases quickly when the characteristic size of the dynamic magnetic non-uniformities (the wavelength of a spin wave in the case of a continuous film) decreases. The latter size scales with the size of the element, and so the exchange interaction can not be neglected anymore in the case of nanoscale elements. Moreover, a proper account of the exchange interaction can be essential even in micron sized elements when spin wave modes are confined in the regions of non-uniform demagnetizing field near the element edges that are much smaller than the elements (so called edge[19] or spin wave well modes[25,26]).

As the size of a magnetic element is reduced, its morphology becomes increasingly important. In particular, the edge roughness and the boundary conditions at the perimeter of the elements play the major role in determining the frequency of the edge modes. Thus, the concept that the elements can be described as uniform prisms with sharp edges should also be expected to fail at the nanoscale, although it worked well in many studies of magnetization dynamics in micron sized elements.

In this chapter, we will present results of studies of precessional magnetization dynamics in arrays of magnetic nano-elements in which the exchange interaction, the edge roughness, and the breakdown of the translational symmetry play a key role in the observed phenomena.[19,27–30] This is in contrast to previous works devoted to investigation of magnetic nano-elements that are nominally sub-micron in size but can be reasonably well described using concepts developed for micron sized magnetic elements. The development of theoretical understanding of the underlying physics is still in progress.[31–34] Therefore, we used numerical algorithms to describe observed phenomena, in particular the Object Oriented Micromagnetic Framework (OOMMF).[35] The latter software has been developed at the National Institute of Standards and Technology (NIST) and successfully applied world-wide to simulate high frequency magnetization dynamics in thin film magnetic elements in real space and time.

Specifically, we used time resolved scanning Kerr microscopy[36–38] (TRSKM) as a submicron probe of the magnetization dynamics at the centre of an array of identical magnetic nanoelements, recording its time dependent response to a pulsed magnetic field. The frequencies of the precessional modes of the sample were determined from fast Fourier transform (FFT) spectra calculated from the time resolved signals acquired for different values and orientations of the bias magnetic field. OOMMF simulations were then used to reproduce the observed variations of the mode frequencies, and, once the agreement between the experiment and the simulations was established, to map the spatial

Table 1. The dimensions of the elements and the element arrays studied in this chapter are summarized.

Element size (nm)	Element aspect ratio (size to thickness)	Array size (μm)	Number of elements in array	Edge-to-edge separation (nm)	Array filling factor (%)
6000	2410	N/A	1	N/A	100
630	253	3.3	5×5	37.5	91
425	171	4	9×9	21.9	91
220	88	4	13×13	95	51
120	48	3.9	25×25	37.5	59
64	26	3.66	33×33	48.4	33

distributions of the out-of-plane component of the dynamic magnetization and to calculate the associated effective fields.

2. Samples and Experimental Technique

A square element of 6 μm size and square arrays of about 4 μm total size with 64, 120, 220, 425, and 630 nm elements were patterned using electron beam lithography. The dimensions of the elements and the element arrays are summarized in Table 1. Scanning electron microscope (SEM) images of some elements are presented in Figure 1. In particular, one can see that the 120 nm elements are slightly rounded, and the 64 nm elements are much more rounded. The nominal com-position of the elements was Si/Ta (50 Å)/$Co_{80}Fe_{20}$ (10 Å)/$Ni_{88}Fe_{12}$ (27 Å)/Ta (100 Å), which is similar to that used in actual spin or tunnel valve magnetic field sensors. The values of the thicknesses were traceable to calibration measurements performed by X-ray diffraction. A comparison of the results of vibrating sample magnetometry (VSM) measurements made upon sheet samples of different thickness co-deposited with the elements showed that 12.1 Å of $Ni_{88}Fe_{12}$ was lost due to interdiffusion with Ta, leaving a $Ni_{88}Fe_{12}$ layer with thickness of 14.9 Å. The compositions of the $Ni_{88}Fe_{12}$ and $Co_{80}Fe_{20}$ alloys were chosen so that the magnetostriction of these two layers would compensate each other, and the bilayer as a whole would have a small negative magnetostriction.[39] The $Co_{80}Fe_{20}$ alloy composition had a stable fcc structure. The direction of the easy axis was set in the samples by field annealing. The magnetizations of the $Co_{80}Fe_{20}$ and $Ni_{88}Fe_{12}$ were deduced to be equal to 1445 emu/cm^3 and 585 emu/cm^3, respectively. Since the two layers were strongly coupled by the interlayer exchange interaction, we treated them as a single layer with thickness equal to the sum of their thicknesses (24.9 Å), and with magnetization given by an arithmetical average of their saturation magnetizations (930 emu/cm^3). The uniaxial and average surface anisotropy constants K_2 and K_s were found to be equal to 5120 erg/cm^3 and 0.11 erg/cm^2, respectively. In order to make pump-probe measurements, the elements were formed in between the tracks of a Au transmission line structure with a 30 μm track width and separation, which is shown in Figure 2, so as to experience an out-of-plane pulsed magnetic field. The measurements were performed in the TRSKM configuration at a wavelength of 790 nm. In this configuration, the measured signal was proportional to the polar Kerr rotation, and hence, to the out-of-plane component of the magnetization. In measurements made upon the element arrays, several elements were typically in the area probed by the microscope. Hence, the measurement was sensitive only to the average response of the elements falling under the probe spot. The maximum pump-probe time delay was 4 ns. An interdigitated photoconductive switch made from Au on a GaAs substrate was connected to the transmission line structure, and used for generation of the pulsed magnetic field. The optically gated pulsed field had a rise time of about 40 ps and a decay time of about 2 ns. A static field **H** was applied in the plane of the sample and its strength and orientation were varied during the experiment. Each value of the static field was approached from a much greater value, typically about 1–2 kOe, that was sufficient to cause saturation. The orientation of the bias field will be denoted by the angle Θ made with an axis perpendicular to the tracks so that Θ = 90°, 270° and Θ = 0°, 180° when the field is parallel and perpendicular to the tracks, respectively.

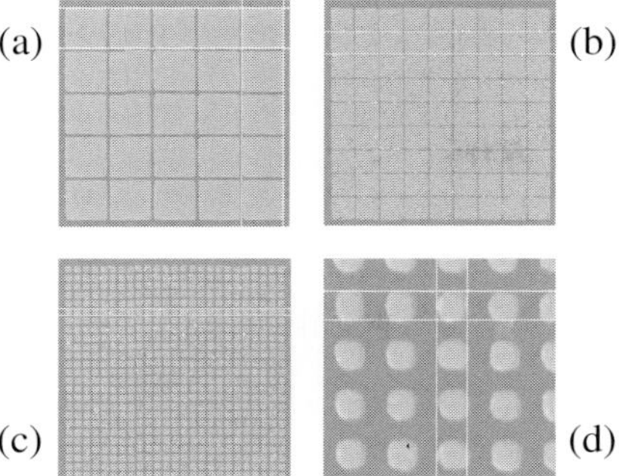

Fig. 1. Scanning electron microscope images of the arrays studied are shown for elements of (a) 630 nm, (b) 425 nm, (c) 120 nm, and (d) 64 nm size. The cursor size in (a), (b), and (c) is equal to the corresponding element size, while in (d) it is equal to 65 nm.

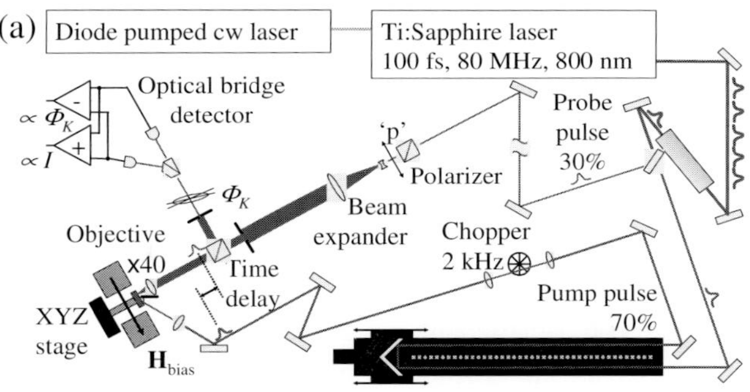

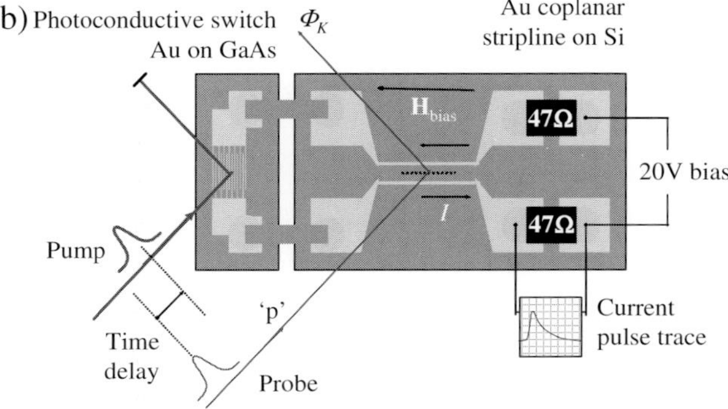

Fig. 2. The setup (a) and the device (b) used in the experiments are shown.

3. Characterization of the Micron Sized Element

First, the measurements made upon the 6 μm element are described. Two types of dynamic Kerr measurement were performed with the probe spot focused at the centre of the element. Firstly, the static field strength H was varied with the field vector **H** parallel to the expected directions of the easy (EA) and hard (HA) anisotropy axes of the element, which were parallel and perpendicular to the tracks of the transmission line structure, respectively.

Figures 3(a) and 3(b) show typical measurements of the time dependent Kerr rotation. A multi-exponential background due to the pulsed field has been subtracted from the signals. The signals reveal a damped precession. The FFT spectra of the signals in panels (a) and (b) are shown in Figures 3(c) and 3(d), respectively. The oscillations decayed well within the duration of the scan, and so, the FFT was performed using a rectangular window function. Kerr images of the dynamic magnetization distribution acquired at different time delays between the pump and the probe (not shown) did not reveal any spatial nonuniformity. Hence, the only observed mode of magnetization precession was identified as the quasi-uniform mode, as might be expected for such a large and thin element.

In a second type of measurement, the direction of the bias field was varied, while its magnitude was kept constant. In Figure 4, the FFT spectra are presented for different directions of a bias field with magnitude of 240 Oe. The uniform precession frequency was determined from the FFT spectra by fitting the peaks to a Gaussian function. The extracted frequencies are plotted in Figure 5(a) as a function of the bias field magnitude, and in Figure 5(b) as a function of the bias field direction.

The experimental points from both sets of measurements were fitted to a macrospin model for a continuous magnetic film. In the fitting, a quasialignment condition was assumed, i.e. the magnetization was assumed to be parallel to the bias magnetic field. The validity of the assumption of quasi-alignment was verified by performing dynamical

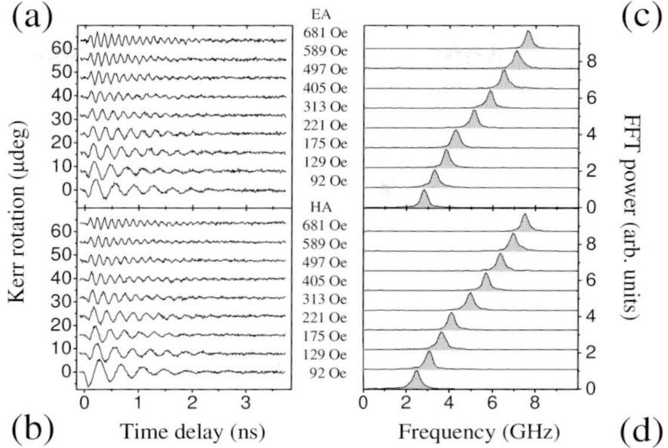

Fig. 3. Typical time resolved Kerr rotation signals measured from the 6000 nm element are presented in panels (a) and (b) for different values of the bias field applied parallel to the EA and HA, respectively. The background due to the pulsed field profile has been subtracted from the signals. The FFT spectra of the corresponding signals in panels (a) and (b) are presented in panels (c) and (d), respectively.

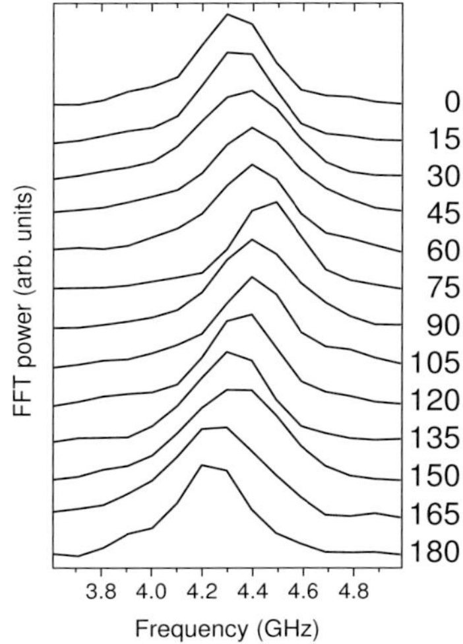

Fig. 4. The FFT power spectra of time resolved signals acquired from the 6 μm element are shown for a bias field of the 240 Oe and the values of angle Θ indicated to the right of the figure.

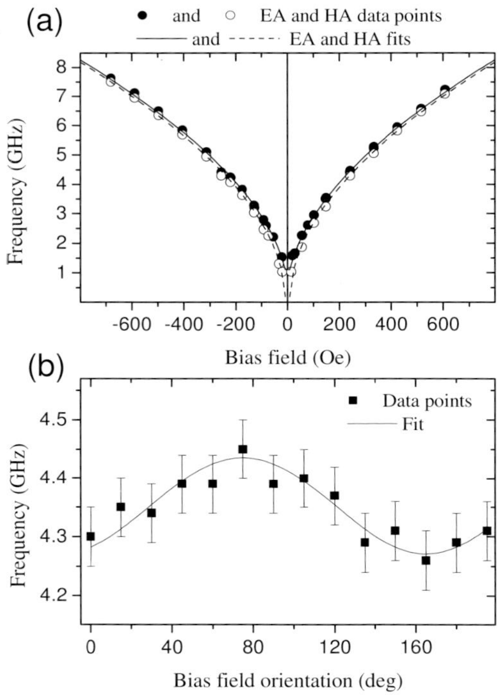

Fig. 5. The dependence of the uniform mode frequency upon the bias field magnitude and orientation is presented for the 6 μm element in panels (a) and (b), respectively. The symbols are data points and the lines are least square fits to the macrospin model equations.

macrospin simulations in which the static orientation of the magnetization was determined using the steepest descent method. The Landé g-factor and the saturation magnetization were fixed at values of 2.1 and 930 emu/cm^3, respectively. From the fitting, it was determined that the uniaxial and surface anisotropy parameters had values of 4140 ± 330 erg/cm^3 and 0.156 ± 0.022 erg/cm^2, which are 19 % lower and 29 % higher respectively than the values obtained from the static magnetometry measurements. The sample EA was canted by about 10° from the expected direction parallel to the tracks of the transmission line structure. However, throughout this chapter we take EA and HA to mean their expected directions, which corresponds to the directions in which the field was actually applied in the experiments. No four-fold anisotropy was observed. In fitting the angular dependence of the mode frequency, the assumed value of the static magnetic field was decreased by 2 Oe, that is, by less than 1%, which is well within the field calibration error.

4. Crossover to Non-Uniform Precession in Magnetic Nanoelements of Reduced Size

The time resolved Kerr signals and their FFT spectra are presented in Figure 6 for the nanoelements of different sizes. The reduced areal coverage led to a weak transient Kerr response with amplitude of just a few micro-degrees, so that a much longer time was required to acquire a signal of reasonable quality. Hence, only measurements for the EA and HA directions and five different magnitudes of the bias field were performed. The 630, 425, 120 and 64 nm element arrays showed a single mode. The time resolved signals acquired from the array of 220 nm elements at the four higher field values exhibit beating, and their FFT spectra suggest the presence of two modes, albeit of reduced intensity. At the field value of 152 Oe, a single mode was observed from the 220 nm elements.

Figure 7 presents the frequencies of the observed modes together with curves obtained by fitting the frequencies of the 6 μm element to the macrospin model equations. The deviation from the latter curves is relatively small for the arrays of 425 and 630 nm elements and much greater for the arrays of 220, 110, and 64 nm elements. The dependence of the mode frequencies upon the element size is presented in Figure 8 for the field applied parallel to the EA direction. One can see that this dependence is complicated and non-monotonic. For the 630 and 425 nm element arrays, the frequency increases somewhat as the element size decreases. In the 120 and 64 nm element arrays, the same trend seems to hold, but the frequencies are reduced in comparison with those in the 630 and 425 nm element arrays. In the 220 nm element array at the four higher field values, the higher frequency modes continue the trend

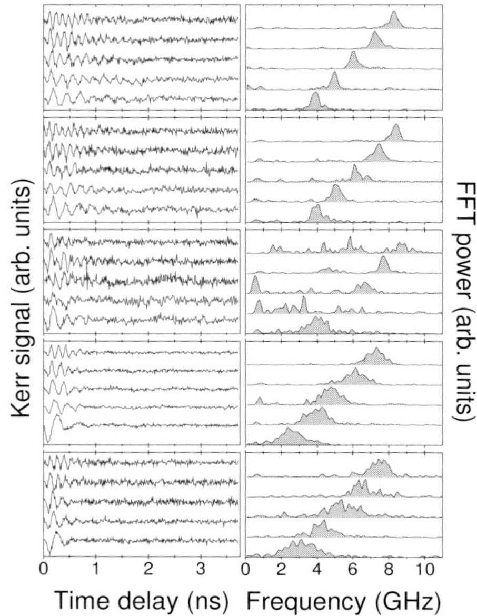

Fig. 6. Time resolved Kerr signals obtained from the element arrays and their FFT spectra are presented in the left and right columns, respectively. In each panel, curves from top to bottom correspond to bias field values of 772, 589, 405, 267 and 152 Oe, respectively. The bias field was applied in the EA direction.

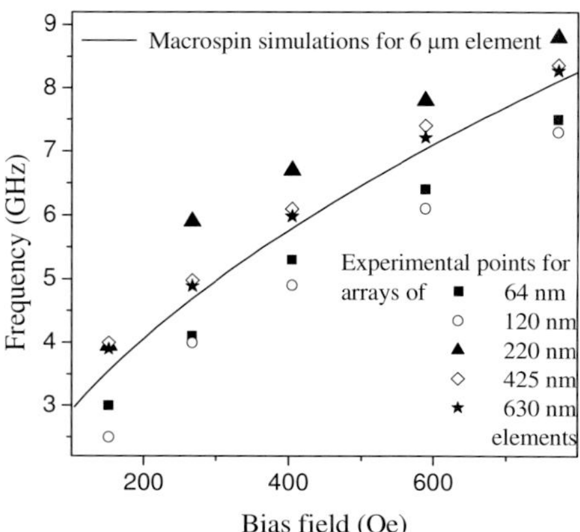

Fig. 7. The mode frequencies extracted from the FFT spectra shown in Figure 6 are plotted as a function of the magnitude of the bias field applied in the EA direction. The lines are the macrospin model fits of the 6 μm element frequencies from Figure 5(a).

observed for the 630 and 425 nm element arrays, while the weaker low frequency modes continue the trend of the 120 and 64 nm element arrays. Hence, we observe two branches in the dependence of the frequency upon element size, with a crossover from one to the other at an element size close to 220 nm. In the following, we refer to them as the "lower" and "higher" branch. The line width of the FFT peaks was noticeably greater in the 120 and 64 nm arrays in comparison with the 630 and 425 nm element arrays.

For comparison, Figure 8 also shows the precession frequency calculated for individual elements with sizes equal to those in the arrays using OOMMF. The cell size was set to be equal to the size of the element, which corresponds to a uniform distribution of the magnetization in the element. The OOMMF implements analytical formulae derived in Ref. 40 for the average demagnetizing tensor components of a uniformly magnetized prism. Hence, each calculated frequency corresponds to a solution of the macrospin model corrected to account for the demagnetizing field. One can see that, although the experimental data follow the calculated frequencies "on average", the detailed dependence on the element size is not properly described. This demonstrates that the origin of the observed behavior lies in the inherent non-uniformity of the static magnetization and effective magnetic field in the elements, which have non-ellipsoidal shape.[41] Therefore, the macrospin model cannot be used to describe the magnetization dynamics in the nanoscale regime. Thus, no further attempts were made to analyze the observed frequencies with the macrospin model.

To identify the origin of the crossover, we simulated the magnetization dynamics within both single isolated elements and model arrays of 3×3 elements so that the center elements of these arrays had the same neighborhood as elements in the measured arrays. Simulations of the complete arrays were not possible due to the computation time required. The square elements had rounded corners, with a radius of curvature that was estimated to be 15 nm from the SEM images. Each array was divided into square cells with height equal to the element thickness, while the lateral cell size was varied to achieve the best compromise between computational time and consistency of the results.

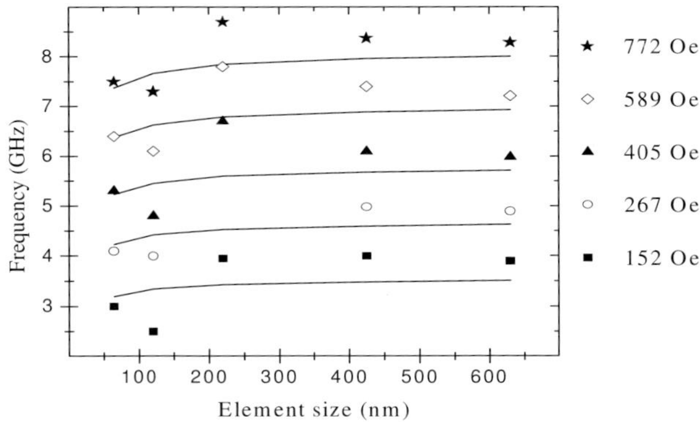

Fig. 8. The mode frequencies extracted from the FFT spectra shown in Figure 6 are plotted as a function of the element size for different values of the bias field. The field is parallel to the EA direction. The symbols represent data points. The lines represent simulated frequencies that were obtained with OOMMF under the assumption that the elements were uniformly magnetized.

In the simulations presented here, a cell size of 7.5, 6, 5, 4, and 2.56 nm was used for 630 and 426, 220, 120, and 64 nm element arrays, respectively. Additional simulations were performed with different cell sizes to ensure that the cell size did not influence the results. The values of the magnetic parameters and the EA direction were assumed to be the same as those obtained from fitting the frequencies measured for the 6 μm element. We assumed a value for the exchange stiffness constant that is 50% greater than that of bulk Permalloy,[29] and a value of 0.015 for the Gilbert damping constant. The surface anisotropy was included as a uniaxial volume anisotropy with the easy axis perpendicular to the element plane. The static field had values of 152, 267, 405, 589, and 772 Oe that were equal to those used in the measurements. The static state at each bias field value was prepared by allowing the magnetization to relax from a uniform state.

In our previous analysis presented in Refs. 19, 27–29 (referred to as "old" in the following), we assumed the bias field to be parallel to an edge of the element. This resulted in significant variation of the magnetic ground state with the value of the bias field. In particular, at field values of 772, 589, and 405 Oe, single elements of all sizes were found to be in the "flower" state.[42] At a field value of 267 Oe, the 120, 220, and 425 nm elements switched to the "S" state,[42] while the 64 and 630 nm elements remained in the flower state. At a field value of 152 Oe, all elements were in the S state. The static state of the center elements in the model arrays was found to be similar in most cases. The exceptions were the 120 and 630 nm elements at a field value of 267 Oe that were in the flower and S states respectively. This transition between the ground magnetic states led to a non-monotonic dependence of the mode frequencies upon the value of the bias magnetic field. In order to ensure that the elements are in the same ground state at all field values, we repeated the analysis with the bias magnetic field canted by 5° from the edge of the element towards the EA. As a result of this, all elements were found to occupy the S state.

The static state was then used as the initial configuration in dynamical simulations, in which an out-of-plane pulsed field with 40 ps rise time, 2 ns decay time, and magnitude of 15 Oe was applied to the sample, as in the experiment. The magnetic state was recorded every 5 ps during the first 5 ns after the excitation. Figure 9 shows the FFT spectra calculated from the out-of-plane component of the dynamic magnetization averaged over the volume of the central element of the different model arrays, while Figure 10 compares the calculated mode frequencies with experiment. Although several modes are generally present in the spectra, their relative amplitudes are different. The variation of the simulated spectra with element size is in qualitative agreement with experiment, and with the simulations presented in Refs. 19 and 29. In particular, the crossover from the higher to the lower frequency branch occurs at an element size somewhere between 425 and 220 nm. At the same time, a detailed comparison of the experiment and simulations reveals some differences. For the element sizes of 630 and 425 nm, a single mode dominates the response at all field values both in the experiment and the old simulations. In the present analysis, this is the case only at the field value of 152 Oe, while the lower and higher branch modes have comparable amplitudes at the higher field values. For 220 nm elements, the lower branch mode also has somewhat greater amplitude than the higher branch. At the element sizes of 120 and 64 nm, the agreement

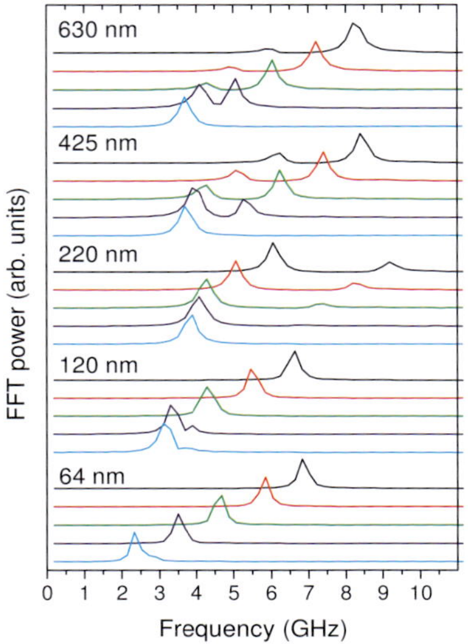

Fig. 9. FFT spectra of the out-of-plane component of the average dynamic magnetization of the center element in the model arrays are shown for different values of the static field. For each element size, curves from top to bottom correspond to bias field values of 772, 589, 405, 267 and 152 Oe, respectively.

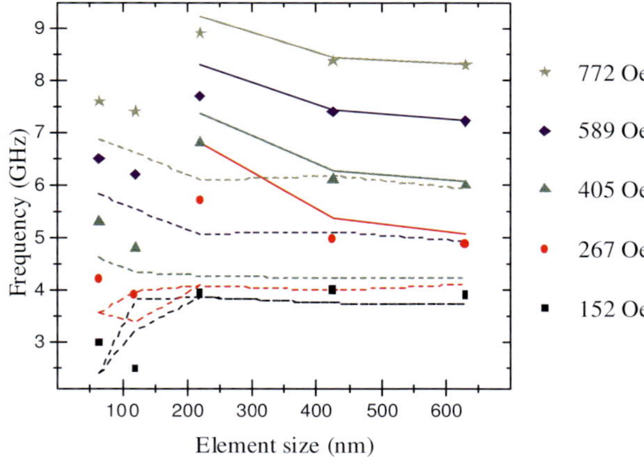

Fig. 10. Extracted mode frequencies are plotted against element size for different bias fields. Symbols represent data points, and the curves are simulations. The solid and dashed lines denote modes ascribed to the higher and lower branches respectively.

with the experiment has improved in the new simulations. In particular, the frequency varies monotonically with the field value. Quantitatively, the agreement is the similarly good for the 630 and 425 nm elements, has improved for the 120 and 64 nm elements, and has become worse for the higher branch from the 220 nm element. The simulations also reproduce the experimental observation that the frequencies for a particular branch decrease monotonically with increasing element size. Let us remark here that when we say that the response of a sample is dominated by a single mode we mean that the other modes cannot be effectively excited by the uniform pulsed magnetic field of finite bandwidth and detected by optical spots much larger than the element size. For example, in the 64 nm element, the peak amplitude of the other modes was found to be more than three orders of magnitude smaller than that of the dominant one.

In order to study various factors affecting the high frequency response of the nanoscale magnetic elements, we have performed further numerical simulations with the static field applied parallel to the element edge.[29] The FFT spectra calculated from the simulated response of elements of different size using different models are compared in Figure 11 for the bias magnetic field value of 772 Oe.

The static magnetization in non-ellipsoidal elements is non-uniform, but normally lies within the plane of the element. The pulsed field used to reorient the magnetization in real devices is also more likely to be applied in the plane of the element. When the pulsed field is out-of-plane the initial torque acting upon the magnetization is uniform in magnitude but non-uniform in direction. However, when the pulsed field is in plane, the initial torque may also be non-uniform in magnitude. Therefore pulsed fields of different orientation may in principle give rise to modes of different spatial character. In addition, the in-plane pulsed field may affect the value of the mode frequency. However, as seen from Figure 11, the simulations of the response of the same individual perfectly shaped square elements to the in-plane and out-of-plane pulsed fields gave rather similar results with a somewhat increased edge to center mode amplitude ratio in the case of the in-plane excitation. This implies that the non-uniformity of the magnetization dynamics reported

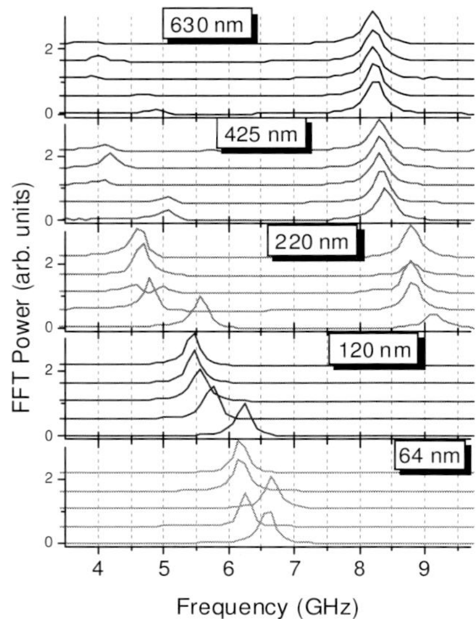

Fig. 11. FFT power spectra of simulated time resolved signals are presented for elements of different size for the field value of 772 Oe. In each set of spectra, curves from top to bottom correspond to simulations for (a) a single perfect square element with out-of-plane pulsed field, (b) a single perfect square element with in-plane pulsed field, (c) a single element of realistic shape with out-of-plane pulsed field, (d) the center element from a 3×3 array of perfect square elements with out-of-plane pulsed field, (e) a single perfect square element with exchange parameter value increased by 50 % and with out-of-plane pulsed field.

above should also be observed when the pulsed field is applied in plane, as is the case in various technological applications.

The shape of real elements is always imperfect. To study effects associated with the edge roughness, we have performed simulations with an out-of-plane pulsed field for individual elements with shape determined from the SEM images of real elements, such as shown in Figure 1. From Figure 11, the element shape is found to have a greater effect upon the mode frequency than the pulsed field orientation. While the center mode frequency was almost unaffected, the edge mode frequency increased as the element shape diverged from ideal, with the largest increase observed for the 64 nm element. The edge mode of the

220 nm element was observed to split into two modes of smaller amplitude.

The effects of the interaction between different elements within an array have been studied by analyzing the response to an out-of-plane field of the center element within the model 3×3 arrays of perfect square elements. From Figure 11, the interaction between elements within the array is seen to affect the edge mode more significantly than the center mode. This is because the stray dipolar fields are non-uniform and fall off very rapidly on the length scale of the element size.

The value of the exchange parameter for the $Co_{80}Fe_{20}$ alloy was unknown and could not be deduced from preliminary measurements of the uniform response of the 6 μm square element. At the same time, Co-Fe alloys are known to have a larger exchange parameter.[43] Here, the effect of the exchange parameter value has been tested by performing simulations with an out-of-plane pulsed field for single perfect square elements with an exchange parameter that has been increased in value by 50 % relative to that of Permalloy. Figure 11 again shows that the increased exchange parameter value has a much more significant impact upon the edge mode, which is explained by its stronger spatial confinement.

Finally, we simulated the response to an out-of-plane field of the center element within a model 3×3 array of realistically shaped 220 nm elements and with the value of the exchange parameter increased by 50%. Although the experimental and numerical results were in somewhat better agreement, some discrepancy remained and could result from two effects. Firstly, the magnetic parameters could have been modified during the patterning process, and this modification could be different for the nano- and micro-scale elements. Indeed, the agreement could be further enhanced if we tuned not only the exchange interaction parameter but also the surface anisotropy, for example. Secondly, the edge profile of the element may be important. For example, the element cross-section could be more trapezoidal rather than rectangular. Micromagnetic simulations performed for a long wire of such cross-section predicted that the edge mode frequency could be increased even further while the center mode frequency remained almost unaffected,[44] which is similar to the effect of using realistic element shapes considered here.

To determine the spatial character of the excited modes, their images were generated from the simulated time resolved data using procedure similar to that from Ref. 38. The magnetization at each time delay was recorded as a vector field map. For each pixel, a time resolved trace was generated from the out-of-plane component of the magnetization and Fourier transformed. The calculated FFTs were then used to reconstruct images of the magnitude and phase of the dynamical magnetization at the frequencies of the modes that were previously identified from the FFT spectra obtained from the average response of the center element. The images of the FFT magnitude are normalized to its maximum value in the center element.

The images of the dynamic magnetization together with those of the static magnetization and total effective magnetic field at the bias magnetic field values of 152, 267, and 405 Oe are plotted in Figures 12–14 for the 630, 220 and 64 nm elements. As in the case of the FFT spectra, at each element size, such images look almost the same for the field values of 405, 589 and 772 Oe, and so images for 589 and 772 Oe are not shown. Also, we do not discuss images for the 425 and 120 nm elements since they are qualitatively similar to corresponding images for the 630 and 120 nm elements, respectively. At all values of the bias magnetic field and of the element size, the static magnetization is in the S configuration. This results in magnetic charges at the edges of the elements. The charges create a large demagnetizing field that peaks near the edges that are perpendicular to the bias field and decays with increasing the distance from them. The images of the total effective field shown in Figures 12–14 demonstrate that the demagnetizing field can not only easily compensate the applied magnetic field but also exceed it locally. The size of such regions in which the total effective field is significantly modified relative to the applied magnetic field (so called "demagnetized regions") increases strongly as the value of the applied magnetic field and/or the element size decreases.

The images of the dynamic magnetization show that the modes from the higher and lower branches have qualitatively different character. In the lower branch, the modes have greater magnitude near the edges that are perpendicular to the bias magnetic field, and so we may say that they

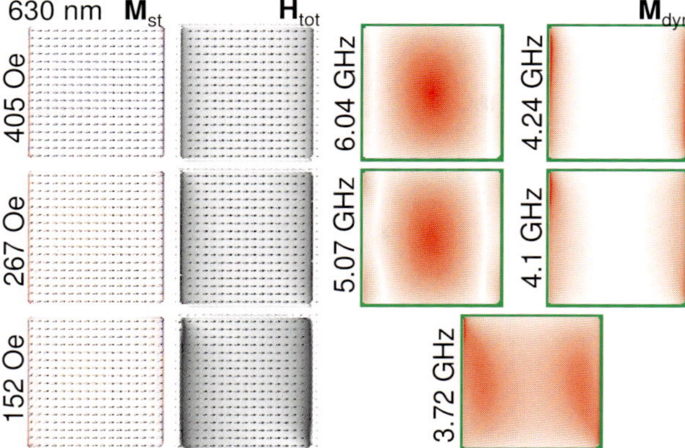

Fig. 12. The spatial configuration of the static magnetization and of the magnitude and direction of the total effective field in the center element of the model arrays is shown in the two left columns for the 630 nm element at the three lower values of the bias magnetic field. In the images of the static magnetization, red and blue depict the density of negative and positive magnetic charges, respectively. In the images of the total effective field, the contrast corresponds to the magnitude of the field with white representing field strengths equal to or greater than that of the bias magnetic field. The two right columns show the spatial distribution of the Fourier magnitude of the dynamic magnetization corresponding to the indicated frequencies. Red and white correspond to maximum and zero magnitudes, respectively. The bias magnetic field is applied at 5° counterclockwise from the horizontal edge of the element.

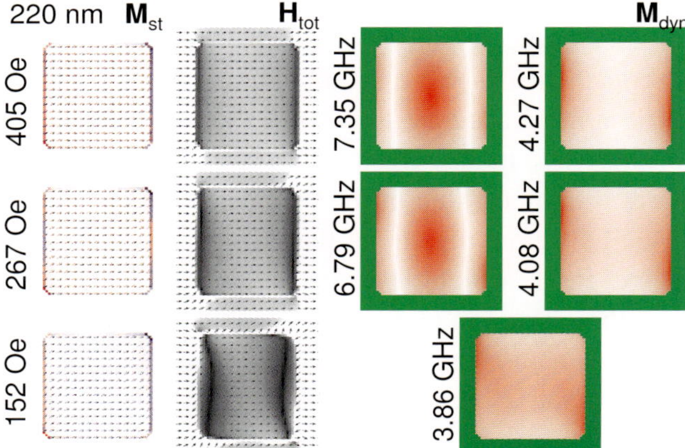

Fig. 13. The simulations of Figure 12 are repeated for the 220 nm element. The contrast and color scheme has the same meaning as in Figure 12.

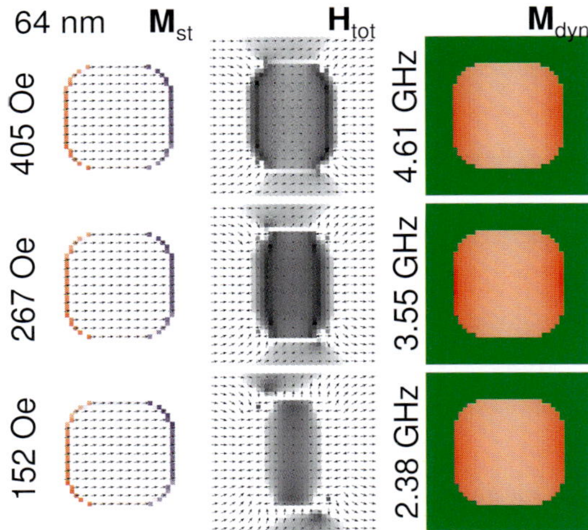

Fig. 14. The simulations of Figure 12 are repeated for the 64 nm element. The contrast and color scheme has the same meaning as in Figure 12.

are localized near the edges. In the higher branch, the modes span the entire element and have two nodal lines at a small distance from the edges. The FFT phase plots (not shown) revealed that the magnetization at opposite sides of the nodal line precesses with opposite phase. Hence, the higher branch modes were called "center modes" in Ref. 19. Here, we see that it is more accurate to refer to them as being de-localized. It is also clear why modes from the two branches have different frequencies at the same value of the bias magnetic field. Indeed, modes associated with different regions of the element see a different "average" total effective field, which leads to a difference in frequency. The crossover of the character of the mode spectrum with reduction of the element size and/or the bias magnetic field can then be explained by the increase of the relative area of the demagnetized region, and hence the relative contribution of the edge mode. The reduction of the total effective field also favors the observation of the low frequency modes, since the magnetization precession has greater amplitude at lower field.

5. Dynamic Configurational Anisotropy in Nanomagnets

The internal magnetic field in elongated or flattened ferromagnetic elements is different when the external magnetic field is applied along the long and short axes. This effect is sometimes referred to as the *shape* anisotropy. Generally, the shape anisotropy is determined by the aspect ratio of the element and exists irrespective of the uniformity of the magnetic configuration. In particular, shape anisotropy occurs in ellipsoidal magnetic elements in which the internal magnetic field and hence the magnetization are uniform. However, a simple calculation shows that the internal magnetic field in the center of a thin square uniformly magnetized element should not depend upon the direction of the external magnetic field used to align the magnetization within the plane of the element.

In magnetic nanoelements with a non-ellipsoidal shape, small deviations of the static magnetization from the uniformly magnetized state result in a strong anisotropy with a symmetry that, to a first approximation, corresponds to that of the shape of the nanoelement.[45,46] Schabes and Bertram identified this so-called *configurational* anisotropy from a numerical study of the dependence of the magnetic hysteresis loop upon the direction of the magnetic field applied to ferromagnetic nanocubes.[45] Cowburn *et al.* studied experimentally the dependence of the total free energy of and the average total effective magnetic field in thin nanoelements upon the direction of a bias magnetic field of constant magnitude.[46] They showed that the effective magnetic field due to the configurational anisotropy could have a magnitude of a few hundred Oersteds in the center of a magnetic nanoelement and so could easily dominate its static magnetic properties.

The manifestations of the configurational anisotropy in the magnetization dynamics of magnetic elements were discussed further in a series of recent publications by different groups. Chérif *et al.* observed a difference between spin wave frequencies within the Brillouin light scattering (BLS) spectra measured with the bias magnetic field applied parallel to the edge and to the diagonal of rectangular elements.[47] The effect of the shape and configurational anisotropies upon the ferromagnetic resonance (FMR) mode was studied by Pardavi-Horvath *et al.*[48] The spatial character of the anisotropic magnetization dynamics in micron sized square magnetic elements was directly studied by TRSKM by Barman *et al.*[49,50] In addition to the fourfold variation of the frequency of the uniform precessional mode, they found that the spatial character of magnetostatic modes of finite wave number depends sensitively upon the direction of the bias magnetic field, and the dephasing of the modes leads to an apparent anisotropic damping of the precessional signal.

In contrast to the isolated element anisotropy discussed above, which can be referred to as an *intrinsic* configurational anisotropy, the effective magnetic field inside an element within an array can also contain an anisotropic contribution due to the stray field from the neighboring elements.[51,52] This anisotropy is mediated by the magnetically unsaturated regions situated near the edges of the elements,[51] and can be referred to as an *extrinsic* configurational anisotropy.

The anisotropic variation of the high frequency magnetization dynamics observed in Refs. 47–52 could be explained by an anisotropic modification of the static effective magnetic field in the magnetic element either due to the configuration of the static magnetization or due to the static stray field from the neighbors. However, Cowburn *et al.* found that the static configurational anisotropy essentially vanishes in squares of thickness equal to or smaller than 3 nm and of size equal to or smaller than 200 nm.[53] Here, we describe our study of the angular variation of the precessional mode frequencies of a square magnetic nanoelement of similar size and thickness. The variation of frequency with bias field orientation is governed in this regime by the variation of both the static *and* dynamic magnetization configuration and the associated dynamic effective magnetic field. There is therefore a *dynamic* configurational anisotropy. The anisotropy is intrinsic and originates from the anisotropy of the magneto-dipole interaction, which has also been known to govern the dispersion of magnetostatic modes in uniformly magnetized magnetic slabs[31] and in magnetic stripes with nonuniform magnetization.[24]

Specifically, we study the magnetization dynamics in the 4×4 μm^2 array of $220 \times 220 \times 2.5$ nm^3 square ferromagnetic elements with a 95 nm edge-to-edge separation, described in the previous section. The time dependent response of the out-of-plane component of the magnetization to a pulsed magnetic field was measured using the TRSKM described before. Again, OOMMF simulations were used to reproduce the observed variations in the mode frequencies, to map the spatial distributions of the out-of-plane component of the dynamic magnetization, and to calculate the associated effective fields.

The measured time resolved Kerr signals and their FFT spectra are presented in Figures 15 and 16 for different values and orientations of the bias magnetic field. In Figure 15, when the field is applied parallel to an element edge, the FFT spectrum contains two modes. As before, the higher and the lower frequency modes are associated with the central area of the element ("de-localized mode") and the element edges perpendicular to the field direction ("edge mode"), respectively. The edge mode FFT peaks are noticeably broader than those of the center mode, which is in agreement with the interpretation in terms of the increased influence of the edge roughness. For the field applied parallel to a diagonal of the element, the FFT spectrum contains only one mode at a frequency that is similar to but slightly smaller than that of the center mode observed when the field was applied parallel to an edge.

In order to elucidate how the transition between the "two mode" and "single mode" regimes occurs, measurements were performed as a bias field of fixed value (589 Oe) was rotated through 360° in steps of 15°. The results are shown in Figure 16. In the FFT spectra obtained with the field aligned 15° from the element edges, a strong peak was observed at a frequency somewhere in between those of the edge and center modes observed with the field parallel to an edge. A smaller less distinct peak was also sometimes observed at a frequency higher than that of the center mode observed with the field parallel to an edge. In the FFT spectra obtained for field directions 15° from an element diagonal, a single peak was observed at a frequency similar to that of the mode observed with the field parallel to a diagonal.

In order to understand the observed behavior, we simulated the magnetization dynamics within both an isolated element and a model array of 3×3 identical elements, using the methods and parameters from the previous section. In order to avoid numerical artifacts due to the

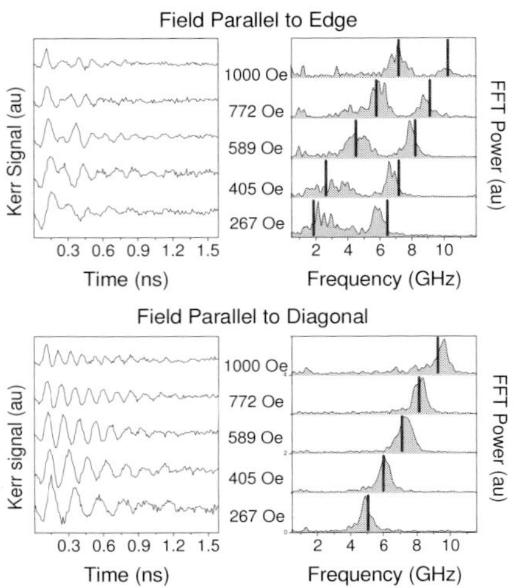

Fig. 15. Time resolved Kerr signals (left panels) and their FFT spectra (right panels) are shown for different values of the bias magnetic field applied parallel to the edge (upper panels) and diagonal (lower panels) of the element. In each panel, curves from top to bottom correspond to bias field values of 1000, 772, 589, 405, and 267 Oe, respectively. The vertical dashes show the mode frequencies from the micromagnetic simulations.

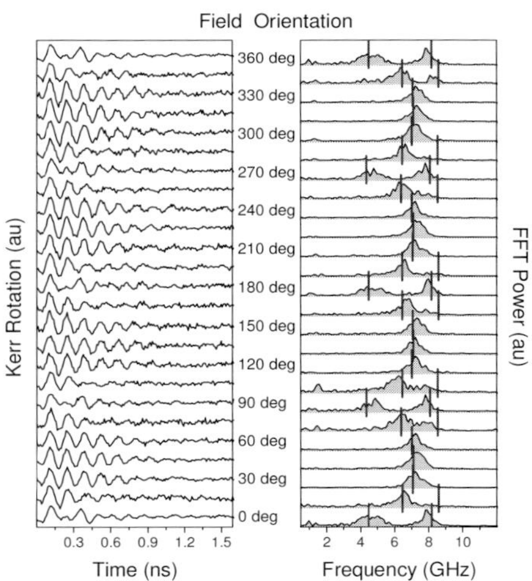

Fig. 16. Time resolved Kerr signals (left panel) and their FFT spectra (right panel) obtained for different orientations of the bias magnetic field are shown for the field value of 589 Oe. For 0°/360°, 90°, 180° and 270° the field lies parallel to an edge of the element. The dashes show the mode frequencies from the micromagnetic simulations.

presence of sharp edges, the shape of the element was derived from an SEM image of the array. At each value and orientation of the bias field, two kinds of dynamical simulation were performed. In the first, the sample was excited by an out-of-plane pulsed field. In the second, a small harmonic out-of-plane field was applied with frequency corresponding to one of the modes identified from the response to the pulsed excitation.

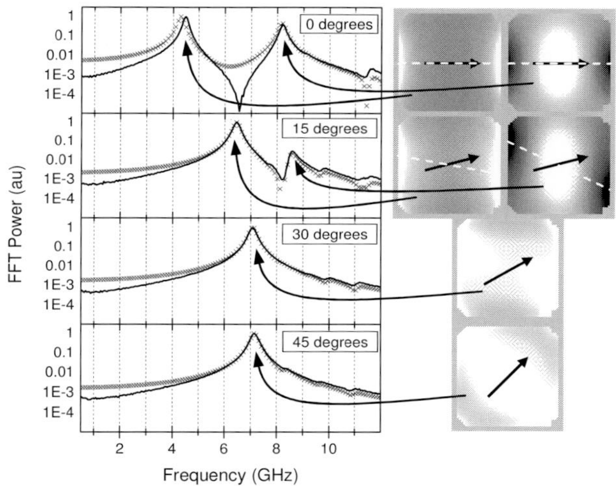

Fig. 17. Simulated FFT spectra of the out-of-plane component of the average dynamic magnetization of the center element in a 3 × 3 array (solid line) and of an isolated element (symbols) in response to a pulsed magnetic field are shown for four different orientations of a bias field of 589 Oe. The dynamic configurations of the magnetization within an isolated element subject to a harmonic magnetic field at a frequency corresponding to the FFT peak marked by the arrow are shown in the images in the right column. White and black correspond to the positive and negative values of the dynamic out-of-plane component of magnetization, respectively. The arrows and the dashed lines in the images represent the directions of the bias field and of the effective wave vector, respectively.

As indicated in Figures 15 and 16, the simulations for the model array with a pulsed magnetic field reproduced well the experimentally observed variation of the mode frequencies with the value and orientation of the bias magnetic field. Figure 17 compares the FFT spectra calculated from the out-of-plane component of the dynamic magnetization averaged over the volume of the isolated element, and of the same element situated in the center of the model array in response to the pulsed excitation. The two simulations were in excellent agreement for all bias field orientations, except for a tiny difference of the edge mode frequency. Therefore we can neglect effects associated with the extrinsic configurational anisotropy in the present discussion. The images of the dynamic magnetization within the individual element in response to the harmonic excitation are shown in Figure 17 as the bias field is rotated between the directions parallel to the edge and to the diagonal of the element.

The mode frequencies calculated from the simulations for the isolated element are plotted in Figure 18(a) for different orientations of the bias magnetic field, while Figure 18(b) shows the corresponding values of the *static* effective field averaged over the center and edge regions of the element, as well as the entire element. As expected from Ref. 46, because of the large aspect (size-to-thickness) ratio, the variation of the static effective field with the applied field angle due to the static configurational anisotropy (< 70 Oe) is too small, on its own, to explain the observed variation of the frequency of the lowest lying mode (> 2.5 GHz). Indeed, one can estimate from Figure 15 that such a frequency change requires a field change of about 250 Oe. In additional simulations, the magnitude of the applied field for the different orientations was slightly adjusted (by less than 16 Oe) so that the value of the average static effective field was the same in each case. This resulted in a modification of the mode frequencies by less than 0.1 GHz. Therefore we conclude that the observed effects must be due to dynamic rather than static configurational anisotropy.

If the Hamiltonian describing a magnetic element is time invariant, the energy of small perturbations to the ground state (magnons) is conserved in the absence of the magnetic damping. The spin wave mode frequencies f then represent "good" quantum numbers, and can be obtained by solving the following eigenvalue problem, derived from the linearised Landau-Lifshitz equation,

$$i2\pi f\, \mathbf{m}_f(\mathbf{r}) = -\gamma\left[\mathbf{m}_f(\mathbf{r}) \times \mathbf{H}_{\mathrm{E}}\left(\mathbf{H}, \mathbf{M}(\mathbf{H}, \mathbf{r})\right)\right] \\ -\gamma\left[\mathbf{M}(\mathbf{H}, \mathbf{r}) \times \mathbf{h}_{\mathrm{E}}\left(\mathbf{m}_f(\mathbf{r})\right)\right] \quad (1)$$

where $\mathbf{M}$ and $\mathbf{H}_{\mathrm{E}}$ are the static magnetization and static total effective field, γ is the gyromagnetic ratio, and $\mathbf{r}$ is the spatial coordinate. The angular variation of $\mathbf{H}_{\mathrm{E}}$ at a specific location is responsible for the static configurational anisotropy. The dynamic effective field $\mathbf{h}_{\mathrm{E}}$ is related to the dynamic magnetization $\mathbf{m}_f$ for a particular mode by a linear integro-differential operator, and is responsible for the dynamic configurational anisotropy.

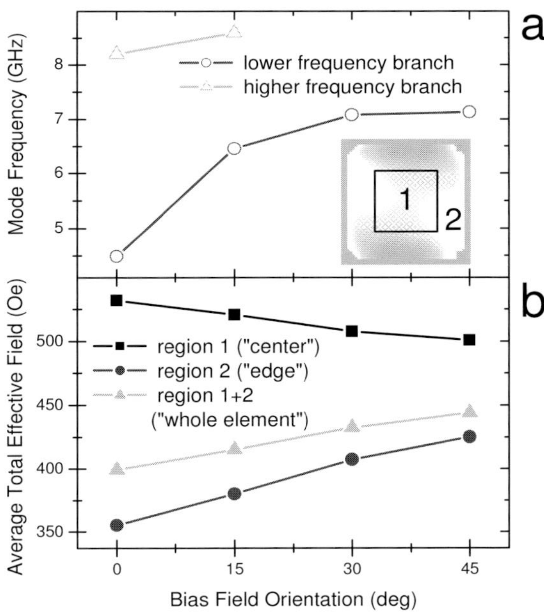

Fig. 18. The dependence of the calculated mode frequencies (a) and of the total effective field averaged over different regions of the sample (b) upon the orientation of the bias magnetic field is shown. The inset shows the center (1) and edge (2) regions of the sample over which the averaging was performed.

In order to gain some further insight into the mechanisms that underlie the observed variations in frequency, it is useful to make comparison with the dispersion of dipolar-exchange spin waves excited within a continuous ultrathin film i.e. in a system with translational invariance for which the in-plane wave vector is well defined. Neglecting any in-plane anisotropy, the frequency of the spin wave modes depends upon the angle φ between the in-plane wave vector ($|\mathbf{k}| = k = 2\pi/\lambda$) and the static magnetization as[54]

$$2\pi f(\varphi) = \gamma\sqrt{\left(H + \frac{2A}{M}k^2 + 2\pi Mkd\sin^2\varphi\right)\left(H + \frac{2A}{M}k^2 + 4\pi M\left(1 - \frac{kd}{2}\right)\right)} \quad (2)$$

where A is the exchange constant and M is the saturation magnetization. Let us now introduce an effective "wave vector direction" as shown in Figure 17. This direction appears to lie parallel to the field when the field is applied along the edge of the element, but seems to be strongly rotated from the direction of the field when the field is applied at 15° from the edge. Equation (2) shows that the dynamic dipolar interaction influences the frequency through both the magnitude of the wave vector (in the terms containing kd) and its direction φ with respect to the static magnetization, while the exchange interaction influences the frequency only through the magnitude of the wave vector. Using the parameter values stated previously, equation (2) can easily account for the observed frequency variation if φ and k are allowed to vary by about 15° and 10% respectively as the orientation of the applied field is varied.

The arguments presented in the previous paragraph may allow us to explain qualitatively the dynamic configurational anisotropy in terms of the dispersion of dipolar-exchange spin waves. A more rigorous analysis requires equation (1) to be solved in terms of a more appropriate set of eigenfunctions. However, the solution should also take into account the transition of the mode character from "edge" to "center" observed in Figure 17 as the orientation of the bias field is changed. As the modes span different areas of the element, they also "see" different static effective magnetic fields and hence different contributions from the static configurational anisotropy, as indicated in Figure 18. However, the frequency of the lowest mode changes only slightly at the mode crossover from the edge to center character observed at the field orientations of 30° and 45°. This suggests that, in the present case, changes in the effective wave vector drive the variation in frequency of the lowest lying mode as the orientation of the static field is changed. Although the regions over which the static field is averaged in Figure 18 provide only a crude approximation of the regions of confinement of the edge and centre modes, they give a good qualitative account of the small variation of the static effective field. A rigorous analytical theory of the observed phenomena has yet to be developed and is beyond the scope of the present discussion.

6. Summary

In summary, the magnetization dynamics of nanoscale magnetic elements of square shape were investigated by means of time resolved optical measurements and micromagnetic simulations. The dominant modes fell on two branches of different frequency. The simulations showed that the magnetization dynamics are nonuniform and that the two branches are associated with centre and edge modes. The frequencies and relative amplitudes of these modes are determined by the spatial configurations of the static and dynamic magnetization and the associated static and dynamic effective fields. The static magnetization and total effective field must be either parallel or anti-parallel to each other so that there is no torque upon the magnetization in the equilibrium state. The spatial configuration of these quantities corresponds to a minimum in the magnetic free energy of an element of a particular size and shape subject to a bias magnetic field of a particular strength and orientation. The static effective field determines the spatial distribution of the dynamic magnetization for a particular mode and thereby the strength of the dynamic effective field and its effect upon the mode frequency. A crossover from a regime in which the modes from the two branches have comparable amplitude to a regime in which a single nonuniform mode dominates the response was observed when the element size and/or the strength of the bias magnetic field were decreased, and when the orientation of the bias magnetic field was rotated towards the diagonal of the elements. The latter effect was described as a dynamic configurational anisotropy. The observed effects may dominate the ultrafast magnetization dynamics of nanomagnets within future magneto-electronic devices.

Acknowledgments

The authors acknowledge the financial support of the UK Engineering and Physical Sciences Research Council (EPSRC) and the New Energy and Industrial Technology Development Organization (NEDO).

References

1. J. R. Eshbach, *Phys. Rev. Lett.* **8**, 357 (1962).
2. M. R. Freeman, R. R. Rur, and R. J. Gambino, *IEEE Trans. Magn.* **27**, 4840 (1991).
3. M. van Kampen, I. L. Soroka, R. Brucas, B. Hjörvarsson, R. Wieser, K. D. Usadel, M. Hanson, O. Kazakova, J. Grabis, H. Zabel, C. Jozsa, and B. Koopmans, *J. Phys. Condens. Matter* **17**, L27 (2005).
4. S. I. Kiselev, J. C. Sankey, I. N. Krivorotov, N. C. Emley, R. J. Schoelkopf, R. A. Buhrman, and D. C. Ralph, *Nature* **425**, 380 (2003).
5. A. I. Akhiezer, V. G. Bar'yakhtar, and S. V. Peletminskii, *Spin Waves* (North-Holland, Amsterdam, 1968).
6. Z. Liu, F. Giesen, X. Zhu, R. D. Sydora, and M. R. Freeman, *Phys. Rev. Lett.* **98**, 087201 (2007).
7. S. O. Demokritov, A. A. Serga, A. Andre, V. E. Demidov, M. P. Kostylev, B. Hillebrands, and A. N. Slavin, *Phys. Rev. Lett.* **94**, 047201 (2004).
8. V. Verrakumar and R. E. Camley, *IEEE Trans. Magn.* **42**, 3318 (2006).
9. J. Podbielski, F. Giesen, and D. Gründler, *Phys. Rev. Lett.* **96**, 167207 (2006).
10. E. A. Vilkov, *Phys. Solid State* **48**, 1754 (2006).
11. A. M. Kosevich, B. A. Ivanov, and A. S. Kovalev, *Phys. Rep.* **194**, 117 (1990), and references therein.
12. R. E. Camley, T. S. Rahman, and D. L. Mills, *Phys. Rev. B* **27**, 261 (1983).
13. P. Grünberg and K. Mika, *Phys. Rev. B* **27**, 2955 (1983).
14. J. O. Vasseur, L. Dobrzynski, B. Djafari-Rouhani, and H. Puszkarski, *Phys. Rev. B* **54**, 1043 (1996).
15. L. Giovannini, F. Montoncello, and F. Nizzoli, *Phys. Rev. B* **75**, 024416 (2007).
16. M. Krawczyk and H. Puszkarski, *Cryst. Res. Techn.* **41**, 547 (2006).
17. V. V. Kruglyak and A. N. Kuchko, *J. Magn. Magn. Mater.* **272-276**, 302 (2004).
18. G. Gubbiotti, M. Conti, G. Carlotti, P. Candeloro, E. Di Fabrizio, K. Y. Guslienko, A. Andre, C. Bayer, and A. N. Slavin, *J. Phys.: Cond. Matter* **16**, 7709 (2004).

19. V. V. Kruglyak, A. Barman, R. J. Hicken, J. R. Childress, and J. A. Katine, *Phys. Rev. B* **71**, 220409 (2005).
20. A. Barman, S. Wang, J. Maas, A. R. Hawkins, S. Kwon, J. Bokor, A. Liddle, and H. Schmidt, *Nano Lett.* **6**, 2939 (2006).
21. A. Barman, S. Wang, J. Maas, A. R. Hawkins, S. Kwon, J. Bokor, A. Liddle, and H. Schmidt, *Appl. Phys. Lett.* **90**, 202504 (2007).
22. E. Noether, *Goett. Nachr. Ges. Wiss.* 235 (1918).
23. G. N. Kakazei, P. E. Wigen, K. Y. Guslienko, V. Novosad, A. N. Slavin, V. O. Golub, N. A. Lesnik, and Y. Otani, *Appl. Phys. Lett.* **85**, 443 (2004).
24. C. Bayer, J. P. Park, H. Wang, M. Yan, C. E. Campbell, and P. A. Crowell, *Phys. Rev. B* **69**, 134401 (2004).
25. J. Jorzick, S. O. Demokritov, B. Hillebrands, M. Bailleul, C. Fermon, K. Y. Guslienko, A. N. Slavin, D. V. Berkov, and N. L. Gorn, *Phys. Rev. Lett.* **88**, 47204 (2002).
26. J. P. Park, P. Eames, D. M. Engebretson, J. Berezovsky, and P. A. Crowell, *Phys. Rev. Lett.* **89**, 277201 (2002).
27. V. V. Kruglyak, PhD Thesis (University of Exeter, 2004).
28. V. V. Kruglyak, A. Barman, R. J. Hicken, J. R. Childress, and J. A. Katine, *J. Appl. Phys.* **97**, 10A706 (2005).
29. V. V. Kruglyak, P. S. Keatley, R. J. Hicken, J. R. Childress, and J. A. Katine, *J. Appl. Phys.* **99**, 08F306 (2006).
30. V. V. Kruglyak, P. S. Keatley, R. J. Hicken, J. R. Childress, and J. A. Katine, *Phys. Rev. B* **75**, 024407 (2007).
31. R. W. Damon and J. R. Eshbach, *J. Phys. Chem. Solids* **19,** 308 (1961).
32. B. A. Kalinikos and A. N. Slavin, *J. Phys. C* **19**, 7013 (1986).
33. P. H. Bryant, J. F. Smyth, S. Schultz, and D. R. Fredkin, *Phys. Rev. B* **47**, 11255 (1993).
34. K. Y. Guslienko, S. O. Demokritov, B. Hillebrands, and A. N. Slavin, *Phys. Rev. B* **66**, 132402 (2002).
35. M. Donahue and D. G. Porter, OOMMF User's guide, Version 1.0, Interagency Report NISTIR 6376, NIST, Gaithersburg, MD, 1999 at http://math.nist.gov/oommf/oommf_cites.html.
36. W. K. Hiebert, G. E. Ballentine, and M. R. Freeman, *Phys. Rev. B* **65**, 140404 (2002).
37. A. Barman, V. V. Kruglyak, R. J. Hicken, A. Kundrotaite, and M. Rahman, *Appl. Phys. Lett.* **82**, 3065 (2003).
38. M. Buess, R. Höllinger, T. Haug, K. Perzlmaier, U. Krey, D. Pescia, M. R. Scheinfein, D. Weiss, and C. H. Back, *Phys. Rev. Lett.* **93**, 77207 (2004).
39. D. A. Nepela and S. Sarhadi, *US Patent* N 5751528 (1998).
40. A. Aharoni, *J. Appl. Phys.* **83**, 3432 (1998).
41. M. Pardavi-Horvath, *J. Appl. Phys.* **95**, 6604 (2004).
42. O. Fruchart and A. Thiaville, *C. R. Physique* **6**, 921 (2006).

43. F. Schreiber and Z. Frait, *Phys. Rev. B* **54**, 6473 (1996).
44. R D. McMichael and B. B. Maranville, *Phys. Rev. B* **74**, 024424 (2006).
45. M. E. Schabes and H. N. Bertram, *J. Appl. Phys.* **64**, 1347 (1988).
46. R. P. Cowburn, A. O. Adeyeye, and M. E. Welland, *Phys. Rev. Lett.* **81**, 5414 (1998).
47. S. M. Chérif, C. Dugautier, J.-F. Hennequin, and P. Moch, *J. Magn. Magn. Mater.* **175**, 228 (1997).
48. M. Pardavi-Horvath, C. A. Ross, and R. D. McMichael, *IEEE Trans. Magn.* **41**, 3601 (2005).
49. A. Barman, V. V. Kruglyak, R. J. Hicken, J. Scott, A. Kundrotaite, and M. Rahman, *J. Appl. Phys.* **95**, 6998 (2004).
50. A. Barman, V. V. Kruglyak, R. J. Hicken, J. M. Rowe, A. Kundrotaite, J. Scott, and M. Rahman, *Phys. Rev. B* **69**, 174426 (2004).
51. C. Mathieu, C. Hartmann, M. Bauer, O. Buettner, S. Riedling, B. Roos, S. O. Demokritov, B. Hillebrands, B. Bartenlian, C. Chappert, D. Decanini, F. Rousseaux, E. Cambril, A. Müller, B. Hoffmann, and U. Hartmann, *Appl. Phys. Lett.* **70**, 2912 (1997).
52. G. N. Kakazei, P. E. Wigen, K. Y. Guslienko, R. W. Chantrell, N. A. Lesnik, V. Metlushko, H. Shima, K. Fukamichi, Y. Otani, and V. Novosad, *J. Appl. Phys.* **93**, 8418 (2003).
53. R. P. Cowburn, D. K. Koltsov, A. O. Adeyeye, and M. E. Welland, *Europhys. Lett.* **48**, 221 (1999).
54. J. F. Cochran, J. Rudd, W. B. Muir, B. Heinrich, and Z. Celinski, *Phys. Rev. B* **42**, 508 (1990).

Chapter 4

MODE STRUCTURE OF FERROMAGNETIC SQUARES

Matthieu Bailleul
Institut de Physique et Chimie des Materiaux de Strasbourg, Strasbourg, France

Christian Back
Institut für Experimentelle Physik, Üniversität Regensburg, Regensburg, Germany

1. Introduction

The detailed understanding of the magnetization configuration of ferromagnetic contacts on length scales of several micrometer or even in the nanometer range of sizes will become important as soon as one would like to use these contacts for example as injection pads for the injection of spin polarized carriers into semiconductors or as small magnetic memory cells as for example for MRAM applications. Micromagnetics will thus become important for two reasons: first the determination and control of the static magnetic ground states is governed by the interplay of dipolar and exchange interactions. Second, the dynamic properties of elements of a particular shape which are also described by micromagnetics will determine the high frequency response of such elements. In this book several authors have considered elements of various shapes and sizes. We add here a compilation of data obtained on square shaped micromagnetic objects in the micrometer size range.

Micromagnetism is usually considered as a paradigm for multi-scale problems; because of the interplay between the long-range dipole interaction, the short-range exchange interaction and the constraint of a constant magnitude for the magnetization vector, a whole zoology of strongly inhomogeneous micromagnetic configurations is available.[23] Amongst those, the most popular object is certainly the domain-wall (DW). These transition regions are generally much narrower (5–100 nm) than the magnetic domains they separate, but they usually dominate the quasi-static behavior of a magnetic particle, basically because they are relatively easy to move. Domain-walls also play a significant role in determining the microwave response of ferromagnets. This was recognized in early studies of the microwave permeability of magnetic materials, where the distinction between low frequency absorption peaks (associated with the motion of domain-walls) and high frequency ones (associated with the resonance of the domains) was made.[39] This picture was refined in the context of magnetic films exhibiting perpendicular magnetic anisotropy which adopt well-defined bubble or stripes domain structures (see e.g. Ref. 36). In particular, the theory of domain-wall resonance was developed to account for the ability of domain-walls to oscillate around their equilibrium position.[30]

Due to the improvements of imaging techniques and broadband microwave techniques, it has recently become possible to address quite directly the dynamics of multi-domain configurations in the context of soft thin-film elements.[7,8,28,33–35,40,43] The model system for those studies is the Landau configuration (see Fig. 1). This highly symmetrical configuration is indeed the most stable remanent state for large enough thin square element. The purpose of this chapter will be to summarize the recent findings about the normal modes of the magnetization field in the Landau configuration. A brief description of the equilibrium state is first provided in Section 2. The experimental methods and results are then described in Sections 3 and 4. The micromagnetic simulations used to reproduce these results are described in Section 5 and a theoretical interpretation of the modes is finally provided in Section 6.

2. The Structure of the Landau State

2.1. *Equilibrium magnetic configuration*

At the most basic level, it is the long-range dipole interaction which is responsible for the formation of domains in ferromagnets. As recognized by Landau,[29] the magnetization tries to arrange itself so as to avoid the building of large dipole fields, which suggests a general flux-closure rule. In the context of soft thin film elements, configurations providing zero dipole field are obtained with the help of a simple geometrical rule;[23,42] the magnetization is required to lie in the plane (in order to avoid building of surface charges on the film surfaces) and parallel to the closest border of the magnetic particle (so as to avoid building of surface charges on the elements border).[a] For polygonal elements, this rule generates a conflict on lines lying along the bissectors of each corner. These lines are interpreted as domain-walls. Following this rule for a square element, we end up with the Landau configuration displayed in Fig. 1(a): the element decomposes into four domains having the shape of iso-rectangle triangles magnetized along their hypothenuse. These domains are separated by four 90° domain walls running along the diagonals of the square.

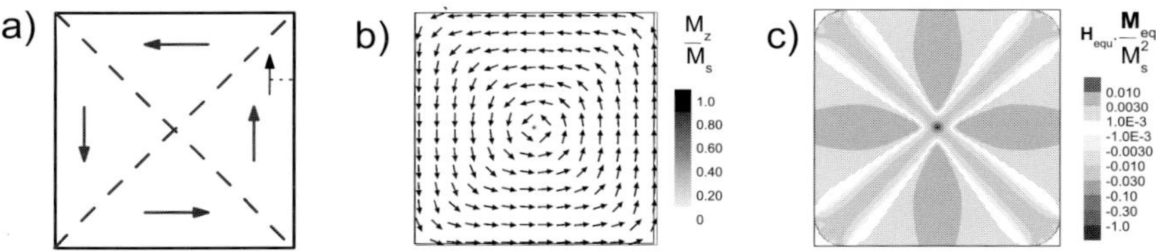

Fig. 1. The equilibrium Landau configuration for a square dot. (a) Four domains — four domain-walls configuration obtained with the help of the Van den Berg construction.[42] (b) Simulated equilibrium magnetization distribution for a $w = 1$ μm wide and $t = 10$ nm thick dot (arrows). The gray scale shows the out of plane component of the magnetization. The material parameters correspond to iron rich permalloy (saturation magnetization $\mu_0 M_s = 1.13$ T, exchange constant $A = 10^{-11}$ J.m^{-1}, negligible magnetocristalline anisotropy). Each corner is rounded with a radius of curvature of 100 nm (typical resolution for electron-beam lithography). The simulation was conducted with the help of the OOMMF code[2] using $2.5 \times 2.5 \times 10$ nm ^{3} cells and starting from a four domain configuration. (c) Simulated equilibrium field profile. A double logarithmic color scale is used to account for the very different values observed in the different parts of the domain structure.

This purely dipolar description should be refined by including also the short-range exchange interaction which forbids any abrupt change of the magnetization. To our knowledge, there exists no analytical formula for a realistic description of such a domain structure. As a consequence, we resort to a numerical calculation obtained with the help of a standard micromagnetic simulation code (see Section 5 for a description of the procedure). In the obtained equilibrium magnetization distribution [see Fig. 1(b)], we

[a]Such a construction rule may actually be understood as an asymptotic thin film — large element limit of micromagnetism.[12]

recognize that (i) the domain-walls have acquired a finite width, (ii) a small structure called vortex has developed in the dot center. Indeed, in a small area at the junction of the four domain the magnetization tilts out of the film plane, which is apparent as a very small dark dot in Fig. 1(b) (see the chapter by G. Gubbiotti *et al.* for a description of vortices in circular dots).

To understand more deeply the structure of the domain-walls, we have extracted a one-dimensional profile of the magnetization across one DW. For this purpose, we define two coordinate axis $O\xi$ and $O\tau$ originating at the middle of the bottom right DW and directed respectively across and along this DW [see Fig. 9(d)]. We then plot in Fig. 9(a) M_τ (i.e. the component of the magnetization oriented parallel to the DW) as a function of ξ (i.e. the spatial coordinate across the domain-wall). At the element border, the magnetization is expected to lie parallel to this border, so that $M_\tau(\pm w/2\sqrt{2}, 0) = \pm\sin(45°)$. The magnetization changes between these two values in three steps : for $\xi < -\delta_{DW}$ and $\xi > \delta_{DW}$, the magnetization rotates only gradually while it rotates pretty fast in region $-\delta_{DW} < \xi < \delta_{DW}$ (Here δ_{DW} is the domain-wall core width). In the spirit of the classical core-tail description of Néel walls in thin films,[23,37] this two-scale behavior should be understood by examining the hierarchy of the magnetic interactions: On the shortest scale, it is the exchange interaction which is limiting. As a consequence, a significant part of the magnetization rotation is required to take place within a few exchange length. This short scale magnetization rotation creates in turn a strong demagnetizing field which is quite localized. The core distribution is the result of the interplay between the short-range exchange interaction and this mostly local dipole field. Now, this core also creates a small long-range dipole field and the magnetization in the tail region arranges itself in order to compensate at least partially for this radiated field.[b]

2.2. *Possible normal modes: Energy scales and symmetry considerations*

A basic approach for understanding the dynamics of the Landau state consists in examining the distribution of the equilibrium effective field.

[b]In our case, the direction of the magnetization in the domains is set by the element border via the non-local dipolar interaction. This is in contrast with classical Néel walls encountered in extended thin films, where the direction of the magnetization in the domains is set by a small in plane anisotropy together with a small transverse applied field. Only in the former case can the problem be made purely one-dimensional and solved analytically.[37]

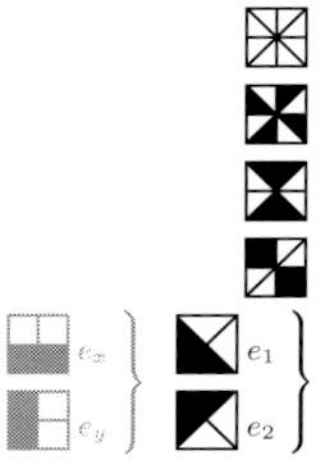

$C_{4\nu}$	E	C_2	C_4^+	C_4^-	σ_1	σ_2	σ_x	σ_y
A_1	1	1	1	1	1	1	1	1
A_2	1	1	1	1	−1	−1	−1	−1
B_1	1	1	−1	−1	−1	−1	1	1
B_2	1	1	−1	−1	1	1	−1	−1
E	$\begin{pmatrix} 1 & 0 \\ 0 & 1 \end{pmatrix}$	$\begin{pmatrix} -1 & 0 \\ 0 & -1 \end{pmatrix}$	$\begin{pmatrix} 0 & -1 \\ 1 & 0 \end{pmatrix}$	$\begin{pmatrix} 0 & 1 \\ -1 & 0 \end{pmatrix}$	$\begin{pmatrix} 1 & 0 \\ 0 & -1 \end{pmatrix}$	$\begin{pmatrix} -1 & 0 \\ 0 & 1 \end{pmatrix}$	$\begin{pmatrix} 0 & 1 \\ 1 & 0 \end{pmatrix}$	$\begin{pmatrix} 0 & -1 \\ -1 & 0 \end{pmatrix}$

e_x e_y e_1 e_2

Fig. 2. Extended character table for the symmetry group $C_{4\nu}$. For the vectorial representation E, the matrices representing the symmetry operations are written for the basis (e_1, e_2) [instead of the equivalent basis (e_x, e_y) sketched in gray]. (Taken from Ref. 4.)

Following the picture of spin wave wells (see the chapter by S. O. Demokritov and V. E. Demidov), this equilibrium effective field can indeed be seen as a potential landscape governing the propagation and/or the confinement of the spin waves. Figure 1(c) shows the quantity $\mathbf{H}_{equ} \cdot \mathbf{M}_{equ}/M_s^2$ corresponding to the magnetization distribution of Fig. 1(b). The product $\mathbf{H}_{equ} \cdot \mathbf{M}_{equ}$ has been chosen because it allows to distinguish between areas where the equilibrium effective field is parallel or antiparallel to the equilibrium magnetization. Moreover, in the presence of dipole and exchange interactions only, this quantity is simply proportional to the magnetic energy density e ($e = -1/2\mathbf{H}_{equ} \cdot \mathbf{M}_{equ}$).[23] Micromagnetic singularities such as vortices and domain-walls concentrate the magnetic energy. They should therefore appear as region with large negative values of $\mathbf{H}_{equ} \cdot \mathbf{M}_{equ}$, which is confirmed in Fig. 1(c). This picture clearly suggests a hierarchy for the normal modes;[35] the lowest frequency mode is expected to be localized where the potential is the lowest (i.e. in the vortex core area), modes lying at slightly higher frequencies are expected to be localized in the domain-wall cores (where the potential is intermediate) while all higher frequency modes would extend also into the domains where the potential is the highest.

Let us now examine how one can use the symmetry properties of the Landau state to simplify the determination of its eigenmodes. For this purpose, we shall use the vocabulary of group theory, as proposed by Yan *et al.*[45]

The symmetry group of a 2-dimensional square dot, $C_{4\nu}$, is sketched in the top row of Fig. 2. It consists of eight operations: the four rotations about the dot center with angles 0, $\pi/2$, π and $3\pi/2$ and the four mirror symmetries with respect to the two diagonals and to the two medians of the square. The in-plane magnetization distribution of the Landau state has the same symmetry as the square itself (as the magnetization is a pseudo-vector, it is required to lie perpendicular to each mirror symmetry line.)[c] Following the Wigner-Eckart theorem, each normal mode of this system is required to belong to one row of one of the irreducible representations (*irrep.*) of this symmetry group.[27] These *irreps.* are listed in the left column of Fig. 2: there are four scalar *irreps.* (the first of which is the identity labeled $A1$) and one vectorial *irrep.* labeled E. The way each *irrep.* transforms under each symmetry operation is indicated by the character table (main body of Fig. 2). It is also shown pictorially on the left side of the table, where the black and white contrast indicates a change of sign enforced by the *irrep.*[d]

The first interest of this symmetry analysis is to classify the modes.[45] But it can also be used to simplify the calculations: As the mode profile in the full square may be deduced from that in one elementary "dissimetrized" triangle (an octant for scalar *irreps.* and a quadrant for the vectorial *irrep.*), one can restrict the eigenvalue problem (i.e the diagonalization) to the dissymetrized body, which reduces significantly the calculation time. This analysis may finally be used to predict which modes are likely to couple to a given excitation (e.g. the small pulse or microwave magnetic field used in the experiment). For this purpose, one determines first which *irrep.*(*s*) the excitation belongs to. For example, a uniform out-of-plane field h_{1z} is associated to the irrep A_1 while a uniform in-plane field h_{1x} belongs to the irrep E (basically because of the precession $M \times h_1$ which causes the magnetization in the left (resp. right) domain to tilt out (resp. into) the plane of the film.). Simulations of the dynamics following the application of pulse fields belonging to the *irreps.* A_1, A_2 and B_1 have been published by Bolte *et al.* (see respectively Figs. 1(d), 1(b) and 1(c) of Ref. 6).

Once we have predicted some very generic features of the normal modes in the Landau state, let us review the experimental results. We shall later come back to theory for a more quantitative and specific description.

---

[c]Note that the out-of-plane vortex core does not behave properly with respect to mirror symmetries. The symmetry group for the full magnetization distribution, labeled C_4, contains indeed only the four rotations.[45] Our symmetry analysis is expected to be realistic for domain and domain-wall modes as long as they do not couple significantly to the vortex mode.

[d]We show in this character table the way a scalar quantity transforms under the symmetry operations. For simplifying purpose, we can assume that the out-of-plane dynamic magnetization also behaves as a scalar (taking into account its pseudo-vector nature would simply result in interverting the characters of *irreps.* $A_1 - A_2$ and $B_1 - B_2$).

3. Experimental Methods

In general, magnetization dynamics of uniform or non-uniform magnetic elements may be studied using a large variety of experimental methods. Here, we focus on the investigation of *non uniformly* magnetized magnetic elements in the micron range of sizes. To obtain a deeper understanding of the dynamic mode structure of such kinds of elements special requirements have to be met by the experimental methods. First it is evident, that magnetization dynamics have to be explored in small or zero applied field to ensure that he specific domain structure is preserved. Second, to allow for a correlation between experimentally obtained results and results obtained by micromagnetic simulations or analytical calculations it is of utmost importance to *image* the dynamic response of individual elements. It thus becomes immediately clear that high resolution imaging techniques are required as the spatial extent of some of the features under investigation is in the sub-micron range. However, some inductive techniques offer a high signal-to-noise ratio and are thus often used to investigate the dynamic response of ferromagnetic elements. In comparison with micromagnetic simulations it then becomes possible to disentangle the complicated mode spectrum.

In this section we describe in details spatially resolved time-domain optical methods (time resolved Kerr microscopy (TR-MOKE), time resolved transmission X-ray microscopy (TR-TXM) and time resolved X-ray photoemission microscopy (TR-XPEEM). Microfocus Brillouin Light Scattering (Micro-BLS)) and vector network analyzer ferromagnetic resonance (VNA-FMR) will only be introduced because they are described in detail in other chapters of this book.

3.1. *Time resolved Kerr microscopy*

One of the most commonly used experimental approaches has been time resolved Kerr microscopy (see Ref. 22 and the chapter by R. J. Hicken *et al.*). It offers the advantage of a fairly high optical resolution of 300 nm in combination with high temporal resolution limited in principle only by the laser pulse width of modern femtosecond laser systems, typically 150 fs. The advantage of this method is that it is table top and can be based on standard laser systems and standard polarization preserving optical microscopy. Usually TR-MOKE is performed as a stroboscopic experiment: the probing laser pulse is phase locked to the source of the magnetic excitation and the answer of the magnetic system is recorded at a fixed phase. To obtain a reasonable signal to noise ratio (SNR), the signal is averaged over thousands of pump-probe cycles. The most frequently used pulsed light source is a commercial Ti:Sapphire laser system producing laser pulses at a central wavelength of 800 nm with a repetition rate of 80 MHz. The pulse width of standard laser sources ranges between 50–200 fs. By frequency doubling using a BBO crystal, laser pulses shorter than 200 fs at a wavelength of 400 nm are obtained. The 400 nm beam is typically used as the probe beam and is consequently directed into a polarization conserving microscope objective lens with a high numerical aperture for maximum spatial resolution. The spatial resolution is limited around 250–300 nm. The reflected or transmitted light is analyzed according to its polarization state in a bridge photo detector situated behind a Wollaston prism. To excite the sample two different modi can be used: (i) a voltage/current pulse can be generated using a short rise time pulse generator which is triggered by a laser pulse from the same oscillator or by launching a current pulse by closing an optical switch. In both cases, the current pulses are launched into a high frequency wave guide and produce the required magnetic tipping field pulse at the location of the sample. This excitation is broad band and the magnetic field contains all frequency components which are present in the Fourier transform of the field pulse. Note, that the relative strength of the Fourier components of the tipping field pulse determines the relative strength of the magnetic modes which are excited in the magnetic element. (ii) The laser oscillator can be synchronized to an rf generator. In this case, a pure sinusoidal waveform containing only a single frequency can be coupled into the waveguide. The probing laser pulses now arrive at a fixed phase with respect to the sinusoidal excitation. In both excitation schemes, the temporal resolution is achieved by moving a motorized mechanical delay stage and by delaying the pump beam with respect to the probe pulse. To enhance the SNR the excitation is gated at some kHz and lock-in detection is used. The sample is mounted on a piezoelectric table which allows to move the sample under the focused beam and to obtain images of the local magnetization by scanning at a fixed time δt after the excitation. The spatial resolution is given by the quality of the beam and the numerical aperture of the objective lens. In our case, we reach $\approx$ 300 nm for the polar Kerr component. It is important to realize that due to the lock-in technique only differences of the Kerr signal with and without excitation ($M_z(t) - M_z(t = 0)$) are detected. The detected Kerr signal is proportional to the *change* of the magnetization due to the excitation.

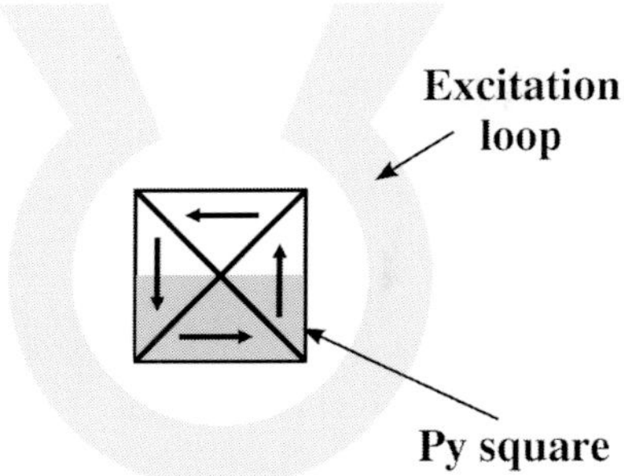

Fig. 3. Sketch of a typical sample layout. The ferromagnetic square in the Landau state is surrounded by a single turn coil which produces a perpendicular field pulse at the position of the sample. The edge length of the square is four micrometer.

A possible way of generating fast, local magnetic field pulses is to launch current pulses into a microcoil, coplanar wave guides or into microstriplines. Figure 3 shows a lithographically defined single turn coil surrounding a flat ferromagnetic square of NiFe. The microcoil is connected to a microstripline which in turn is connected to a current source, in this case an optical switch. We can now record traces in time by placing the probing laser spot somewhere on the sample and scanning the optical delay, or we can fix the delay to a certain time and acquire an image of the response of the sample. By assembling images as a function of time, movies of the response of the magnetization to the excitation may be recorded and analyzed further.

3.2. *Time resolved X-ray microscopy*

A similar principle as in TR-MOKE is used for time resolved X-ray microscopy experiments.[22] Here the probing laser pulse is simply interchanged with an X-ray pulse. Typically the X-ray pulse duration is in the range of 50–70 ps in typical third generation synchrotron sources, in some case it can be reduced to about 10 ps. In the case of transmission X-ray microscopy as discussed in this section, the probing X-rays are focused to a spot of about 30 nm and the transmitted X-ray intensity is recorded in a fast photodiode assembly. The intensity of the X-rays changes due to the X-ray magnetic dichroism (XMCD) effect depending on the alignment of the magnetization with respect to the propagation direction of the circularly polarized X-ray beam. Thus, at fixed helicity the precession of the magnetization can be detected when a large relative change of the projection occurs. The XMCD effect can be as large as several ten percent for typical 3d transition metals, which makes it very attractive for the detection of small changes of the precessing magnetization. In the experiments described below the external stimulus for the magnetization is provided by an rf generator which is phase locked to the X-ray pulses, similar to the set-up described in the previous section.

In a second XMCD based experiment a photoelectron emission microscope (PEEM) is used to detect the changes of the magnetization on a 50 nm length scale.[8,28,35] Here, the electron avalanches produced by the different absorption of photons due to the XMCD effect are detected. A gated detection scheme is used to allow detection of single X-ray bunches which are again phase locked to a laser pulse excitation.

3.3. *Microfocus Brillouin light scattering*

In the chapter by G. Gubbiotti *et al.* and by S. O. Demokritov and V. E. Demidov the working principle of Brillouin Light Scattering (BLS) has been introduced in detail. In the chapter presented here, we show results of BLS measurements using a microfocus BLS set-up with a spatial resolution of about 300 nm.[11] By scanning the sample under investigation using a Piezo scanning stage, the spatial distribution of thermal spinwaves can be detected with great accuracy. Furthermore, it is possible to excite magnetic modes using an rf generator connected to the microstripline.

3.4. *Vector-network analyzer ferromagnetic resonance*

We also show data obtained by Vector-Network Analyzer Ferromagnetic Resonance (VNA-FMR) as described in detail in the chapter of D. Grundler *et al.* We use a vectorial network analyzer (VNA) to perform essentially ferromagnetic resonance (FMR) experiments as a function of externally applied magnetic bias field in-the plane of the squares. VNA-FMR measurements differ from conventional FMR in one important point. Using a VNA one can sweep the frequency at a fixed external bias field while in conventional FMR the exciting rf-frequency is fixed and the external field is swept through the resonance. For micron sized magnetic structures in a domain configuration this means that using VNA-FMR a spin configuration may be conserved in a fixed bias field and the resonance of this configuration is found by sweeping the frequency.

4. Experimental Results

4.1. *In-plane excitation: Gyrotropic mode*

The gyrotropic vortex mode has been studied in detail in ferromagnetic disks (mostly made of permalloy NiFe)).[18] In flat squares in the Landau flux closed state the motion of the vortex core after excitation by an in-plane magnetic field pulse was first reported by Choe *et al.*[8] In their experiment a series of elements (20 nm thick CoFe, edge length between $1 \times 1\ \mu\text{m}^2$ and $2 \times 1\ \mu\text{m}^2$) are placed on top of a Cu waveguide. A laser pulse is used to close an optical switch and a short magnetic field pulse (broad band excitation, magnetic field pulse width about 300 ps) is launched into the waveguide structure. The dynamic response of the elements is imaged using a PEEM. X-ray magnetic circular dichroism (XMCD) at the transition metal L edges probes the direction and size of the element-specific magnetic moment. After the field pulse excitation the vortex is displaced from the center of the element (in the case of Choe *et al.* a displacement of the vortex in the direction of the field pulse is observed[8]). In the subsequent nanoseconds the vortex spirals back towards the center, but does not fully reach its equilibrium before it is excited again by the next field pulse 8 ns later.

The time between excitation has been 16 ns in the experiments of Raabe *et al.*[35] In this case a very similar set-up is used. However, here NiFe elements of larger size are imaged ($6 \times 6\ \mu\text{m}^2$). In their case a rather strong excitation has been chosen. Again PEEM has been used to investigate the response. It is found that the vortex is displace perpendicular to the field pulse excitation and comes to rest after about 16 ns. No signature of the gyrotropic mode is found in these large elements.

In the experiments described above, the position of the vortex is inferred from the images obtained from the domains and the domain walls. In a third synchrotron based experiment van Wayenberge *et al.*[43] and Chou *et al.*[9] have imaged the motion of the vortex directly using transmission X-ray microscopy. Their experiment is based on a quasi resonant single frequency excitation of the vortex. Again the elements are placed on top of a waveguide which is connected to an rf generator. The frequency of the generator is tuned to close to the frequency of the gyrotropic mode (about 500 MHz for a $0.5 \times 0.5\ \mu\text{m}^2$, 40 nm thick NiFe element). In this case the gyrotropic mode can be excited very efficiently. In fact it is easily possible to reach a non-linear regime and even cause switching of the polarization of the vortex core. Moreover, it was possible to image the perpendicular

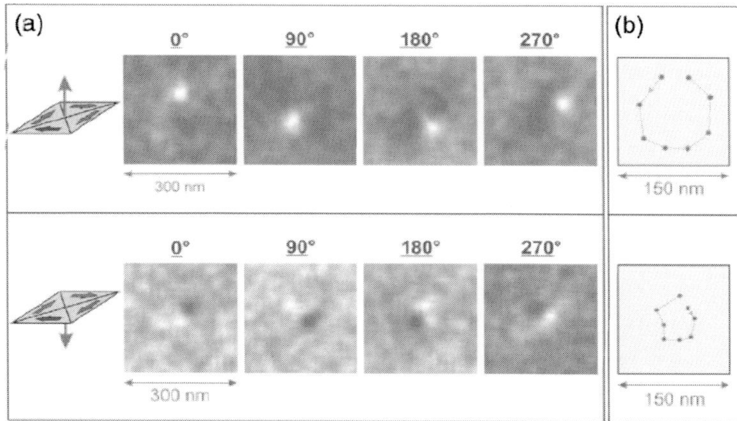

Fig. 4. Time resolved scanning X-ray transmission microscopy image series of a NiFe square with edge length of 500 nm and 40 nm thickness. In both image series (a) the sense of flux closure of the magnetization is clockwise. In the upper panel the vortex core points up, in the lower panel it points down. Images are recorded at phases of 0, 90, 180, and 270 degrees. In (b) the trajectory of the vortex core is shown. It is counterclockwise in the upper panel and clockwise in the lower panel (from Ref. 9).

component of the magnetization in the core of the vortex directly and to follow its trajectory directly, see Fig. 4.

4.2. *In-plane excitation: Domain wall mode*

The domain wall mode was first studied by Park *et al.*[33] In this experiment, an in-plane excitation as described above has been used for 18 nm thick NiFe squares with edge dimension between 1 and 10 μm. After excitation of the elements using a weak (5 Oe) but short (150 ps) in-plane field pulse a time resolved movie has been recorded using TR-MOKE. This movie has been Fourier transformed in each Pixel and has been reassembled to show a movie as a function of frequency. At distinct frequency pronounced features can be evidenced in the spectral images, see Fig. 5. Here, the low frequency response has been attributed to the excitation of the domain walls while the higher frequency response is located in the large triangular domains.

A similar behavior has been found by Raabe *et al.*[35] Because of the much higher resolution of PEEM imaging, the bending of the domain walls corresponding to the first domain wall mode could be imaged directly (see Fig. 6).

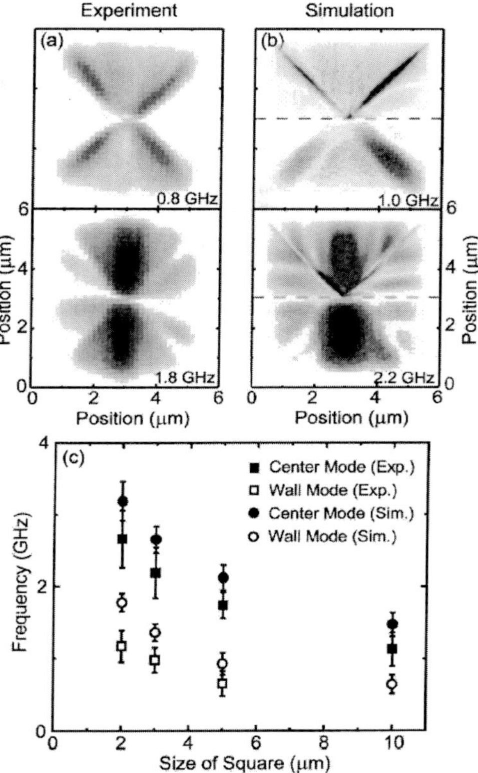

Fig. 5. (a) Comparison of experimental spectral images of the response of a 5 micrometer edge length permalloy square at the two frequencies corresponding to the peaks in the average spectrum. (b) Spectral images obtained from the Landau-Lifshitz-Gilbert. The upper half of each image in (b) shows the unconvolved output of the simulation, while the lower half has been convolved with the optical resolution function. The frequencies shown correspond to the peaks in the simulated spectra, which are slightly higher than their experimental counterparts. (c) Values of the domain center (closed symbols) and domain wall (open symbols) frequencies as a function of square size. Experimental and simulation results are shown by squares and circles,respectively. (Taken from Ref. 33.)

4.3. *In-plane excitation: Magneto-static modes*

In the two experiments described above (Park *et al.*[33] and Raabe *et al.*[35]) a strong excitation could be observed in the large domains. This was attributed to magnetostatic modes propagating within the domains (in agreement with the energy scale considerations of Sec. 2.2) but no detailed

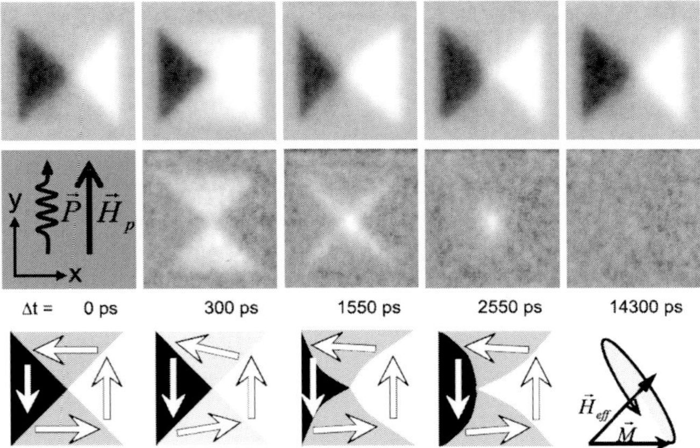

Fig. 6. Selected XMCD images showing the time evolution of the magnetic contrast in 6 μm permalloy squares (top row). Difference images obtained by subtracting the ground state ($\Delta t = 0$) from each images (middle row). With the orientation of the exciting field pulse ($\vec{H}_p = 60\,\mathrm{O}e$, 150 ps rise time) and the polarization $\vec{P}$ sketched in the leftmost image. Sketches showing the prominent features for different delay times are shown on the bottom row. (Data taken from Ref. 35.)

classification of these modes could be given due to the limited experimental resolution. More details were obtained in the out-of-plane excitation experiment reported by Perzlmaier *et al.* (see the next subsection) and in the combined simulation/VNA-FMR experiment described in Section 5.

4.4. *Perpendicular excitation: Magneto-static modes*

In Ref. 34, ferromagnetic squares in the flux closed Landau state have been excited by a perpendicular field pulse. We have investigated squares with edge sizes between 0.75 and 4 μm made out of 16 nm NiFe. In this case the ferromagnetic elements have been placed inside a single turn current loop, see Fig. 3. Launching a current pulse through the loop produces a quasi homogeneous perpendicular magnetic field at the position of the sample. Following the recipe of Park *et al.*[33] a time resolved movie has been recorded using TR-MOKE and several modes have been extracted using Fourier transformation. Exactly the same element has been investigated using microfocus BLS.[11] Microfocus BLS imaging was performed for frequencies in the range from 2 to 8 GHz with a frequency step size of 100 MHz. In order to improve the sensitivity and to filter modes with

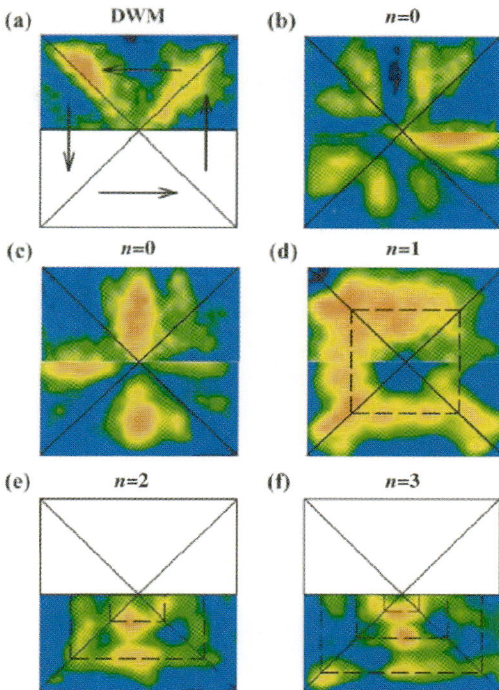

Fig. 7. Distributions of the amplitude of the dynamic magnetization corresponding to the eigenmodes. The upper and lower parts of the images are measured using TR-MOKE and micro-BLS, respectively. Diagonal solid lines indicate the domain walls. Arrows in (a) indicate the directions of the static magnetization. DWM stands for the domain-wall mode. The corresponding transversal quantization numbers n are indicated above the images. (Data taken from Ref. 34.)

fast spatial variations, an external excitation field at a fixed frequency was applied to the sample by applying a monochromatic microwave current through the excitation loop. The probing laser spot was scanned in two dimensions with a step size of 200 nm. the experimental results are presented in Fig. 7. Here we show the combined images obtained by BLS and TR-MOKE for a 16 nm thick NiFe square with 4 μm edge length. At low frequencies (below 1 GHz) we cannot detect any reasonable signal using BLS and consequently the domain wall mode at 0.8 GHz is shown only in the upper half of the images which presents the TR-MOKE data. At intermediate frequencies both methods show the same results. At 2.3 GHz the main (quasi homogeneous) response of the triangular domains is evident. However, at frequencies slightly below this main mode (2.1 GHz) a mode is observable with an additional node in the direction of the static magnetization. This quantization direction is reminiscent of magneto-static backward volume modes (MSBV) in extended thin films (see the chapter by S. O. Demokritov and V. E. Demidov for a description of magnetostatic modes and of their quantization). When going to higher frequencies BLS detection is more advantageous as the high frequency components in the particular field pulse excitation are quite weak.Nevertheless when measuring at 4 GHz, 5.5 GHz and 7.1 GHz modes with nodal lines perpendicular to the static magnetization direction become evident. Due to the geometrical analogy to thin films these modes are labelled Damon-Eshbach (DE) mode. It is expected that at frequencies lower than the main DE modes a zoo of MSBV like modes should be present as these modes exhibit negative dispersion — lowering of the frequency with increasing wave vector — while the DE modes obey a positive dispersion. Due to the limited experimental resolution these modes escape our observation with exception of the one located at 2.1 GHz. The set of domain modes will be described in more details in Section 6.3.

5. Micromagnetic Simulations of the Dynamic Response of Landau Squares

As mentioned earlier, even a simple domain configuration such as the Landau state has a complicated multi-scale nature involving the intrication of dipole and exchange interactions. As a consequence, fully numerical methods are particularly important in interpreting the dynamical properties of multi-domain structures. Such numerical procedures have been developed in the past few years, basically by extending micromagnetic codes originally devoted to the simulation of equilibrium states (see e.g. Ref. 15). In the following, a typical implementation of such a procedure is described. We also mention the limitations od such a procedure and describe how the results obtained can be compared to experimental ones.

On a mesoscopic scale, the equation of motion for the magnetization vector field $\mathbf{M}(r,t)$ is the Landau-Lifshitz equation, which writes:

$$\frac{d\mathbf{M}(\mathbf{r},t)}{dt} = -|\gamma|\mu_0\mathbf{M}(\mathbf{r},t)\times\mathbf{H}_{\mathbf{eff}}(\mathbf{r},t) + \alpha\frac{\mathbf{M}(\mathbf{r},t)}{M_s}\times\frac{d\mathbf{M}(\mathbf{r},t)}{dt}, \quad (1)$$

where $\mathbf{H}_{\mathbf{eff}}$ is an effective field containing all the magnetic interactions. In the case of our soft magnetic dots observed at remanence, it writes:[23]

$$\mathbf{H_{eff}}(\mathbf{r},t) = \mathbf{H_d}(\mathbf{r},t) + \mathbf{H_{exch}}(\mathbf{r},t) + \mathbf{h_1}(\mathbf{r},t), \quad (2)$$

where $\mathbf{H_d}$ is the dipolar field, $\mathbf{H_{exch}}$ is the exchange field and $\mathbf{h_1}$ is a small external time-dependent excitation field (optionally spatially non-uniform). The dipole and exchange field write respectively:

$$\mathbf{H_d}(\mathbf{r}) = (\hat{\hat{\mathbf{G}}} \cdot \mathbf{M})(\mathbf{r}) = \int d\mathbf{r}' \hat{\mathbf{G}}(\mathbf{r},\mathbf{r}')\mathbf{M}(\mathbf{r}'), \quad (3)$$

$$\mathbf{H_{exch}}(\mathbf{r}) = \Lambda^2 \Delta \mathbf{M}(\mathbf{r}), \quad (4)$$

where $G_{\alpha\beta}(\mathbf{r},\mathbf{r}') = -\frac{\partial^2}{\partial x_\alpha \partial x'_\beta}\frac{1}{|\mathbf{r}-\mathbf{r}'|}$ is the kernel of the dipolar interaction,[17] $\Lambda = \sqrt{\frac{2A}{\mu_0 M_s^2}}$ is the exchange length and Δ is the Laplacian operator. Finally, the last term of Eq. (1) is a phenomenological damping term, α being the Gilbert damping constant.

The only approximation in most micromagnetic simulations consist in a space and time discretization. The discretized equation of motion is then used to determine both the equilibrium state and the dynamic response around this state: Starting from a suitable fictitious configuration, one let first the system converge towards a stable state (the external excitation h_1 is set to zero and a high value of damping is usually chosen in this step to accelerate the convergence). Note that the obtained equilibrium magnetization $\mathbf{M}_{equ}$ and effective field $\mathbf{H}_{equ}$ are expected to be parallel or antiparallel because of the equilibrium condition $\mathbf{M} \times \mathbf{H_{eff}} = O$. The dynamics around this equilibrium is then obtained by slightly perturbing this equilibrium (e.g. by applying a small duration and small amplitude pulse field h_1), and registering the subsequent magnetization precession. This data is finally transferred into the frequency domain by Fourier transform.

The results of such a simulation are shown in Fig. 8. This simulation was carried out for interpreting the microwave absorption spectrum measured by VNA-FMR on an assembly of $w = 3$ μm, $t = 30$ nm permalloy squares at remanence [top black curve in Fig. 8(a)].[4] The pulse field h_1 was chosen to be spatially uniform and oriented along x in order to mimic the in-plane microwave field used in the experiment (the latter being produced by a coplanar waveguide located above the dots). The local magnetic susceptibility $\chi_{xx}(\mathbf{r},\omega) = \frac{\widetilde{m}(\mathbf{r},\omega)}{\widetilde{h}_1(\omega)}$ and its spatial average $\langle\chi_{xx}(\omega)\rangle$ were extracted from the Fourier transform $\widetilde{m}$ of the magnetization precession.

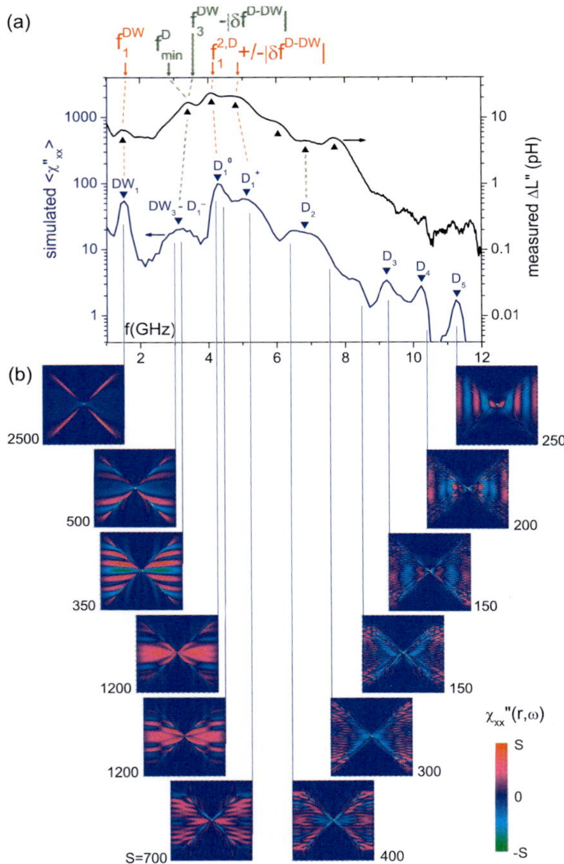

Fig. 8. Microwave spectrum simulated for a $w = 3\ \mu$m, $t = 30$ nm permalloy square in the Landau state. The blue bottom curve in panel (a) shows the simulated spatially averaged susceptibility $\langle \chi_{xx}(\omega) \rangle$. For comparison the absorption spectrum measured with the help of VNA-FMR on an assembly of square dots having the same dimensions is reported as the top black curve. The peaks are named using a convention explained in the text. The arrows above the graph show the frequencies calculated using semi-analytical approaches. Panel (b) shows simulated local susceptibility maps $\chi_{xx}(\mathbf{r}, \omega)$ for twelve selected frequencies. The amplitude S spanned by the color scale is indicated for each map. The material parameters are $\mu_0 M_s = 1.13$ T, $A = 10^{-11}$ J.m^{-1}, $\frac{|\gamma|}{2\pi} = 28$ GHz.T^{-1} and $\alpha = 0.005$. Each corner is rounded with a radius of curvature of 300 nm (typical resolution for UV lithography). The simulation was conducted with the help of the LLG code[1] using $10 \times 10 \times 30$ nm^3 cells and a 2 ps time step. The equilibrium was obtained starting from a four domain state. It was perturbed with a 50 ps, 1 mT, x directed spatially uniform field pulse h_1 and the resulting precession was registered during 17 ns. (Taken from Ref. 4.)

The imaginary part of $\langle \chi_{xx}(\omega) \rangle$ (which represents the power absorption) is shown as the bottom blue in Fig. 8(a). One clearly distinguishes several peaks which reproduce some of the measured one (see Ref. 4 for a detailed comparison). The susceptibility maps reported in Fig. 8(b) allow one to visualize the modes which contribute to those peaks. One recognizes immediately the domain-wall/domain hierarchy predicted in Section 2 and observed in spatially-resolved optical experiments: for low intensity peaks (labeled DW), the absorption is concentrated in or close to the domain-walls while it is more distributed in the domains for higher frequency peaks labeled D. Note that the zero intensity in the top and bottom domains is simply due to the fact that the response function we have chosen vanishes when the equilibrium magnetization is parallel to x. This hierarchy was observed in several other published simulation results.[26,33,45] One also observes in Fig. 8(b) that the nodal structure within the domains changes distinctly as a function of the frequency: counting the antinodes along x, one obtains only one antinode for the three lowest frequency peaks (which are therefore labeled D_1) but one obtains 2 to 5 antinodes for the higher frequency peaks. Counting along y , one obtains about 9 antinodes for the lowest frequency domain peak (D_1^-) but only 3 to 5 antinodes for the intermediate frequency peaks (D_1^0 and D_1^+). Such a domain nodal structure was also observed in the simulations by Bolte *et al.*[6] and Yan *et al.*[45] It is understood qualitatively in terms of quantized spin wave modes (see Section 6.3).

Let us now discuss the limitations of such a procedure. The most obvious one is associated with the spatial discretization itself: any excitation having a wavelength below the cell size is not accessible. For example, the simulation shown in Fig. 8 does not reveal any vortex mode. In order to save computation time, a relatively large cell size (10 nm, i.e. larger than the vortex core diameter) was indeed chosen for this quite large dot. As a consequence, the vortex core did not even nucleate at equilibrium. Simulations carried out on smaller dots (for which a smaller cell size can be chosen) indicate that the fundamental mode is highly localized in the vortex region, in agreement with the hierarchy suggested above.[41,45] The second limitation has to do with the frequency resolution. In order to mimic an experiment, one usually chooses a realistic damping parameter (e.g. $\alpha = 0.005$–0.01 for permalloy), with the consequence that each mode acquires a finite linewidth (typically $\alpha\gamma\mu_0 M_s/2\pi = 150$–$300$ MHz for long wavelength modes in permalloy). Modes whose frequency spacing is lower than this intrinsic linewidth shall naturally mix together, so that they cannot be visualized separately. This is presumably what happens in the spectrum of Fig. 8: the peaks labelled $DW_3 - D_1^-$ to D_2 are much larger than the expected intrinsic linewidth. Moreover, the susceptibility maps extracted at different positions within these peaks are markedly different (compare e.g. the second and third or the fourth and fifth maps), which supports the idea that several distinct normal modes combine together to form these peaks. In order to extend the frequency resolution of this technique, one could use an unrealistic zero damping parameter and calculate the precession for very long times. Alternatively, a different numerical scheme may be used: By linearizing the Landau-Lifshitz equation 1 around equilibrium, writing it into the frequency domain and dropping damping and excitation terms, one obtains a linear integro-differential equation. After discretization this problems reduces to the diagonalization of a so-called "dynamical" matrix. This diagonalization provides all the normal modes of the discretized system, regardless of their symmetry.[14]

In interpreting experiments with the help of such simulations, it is essential to chose a response function accounting for the excitation-detection scheme specific of each experiment. For thermal BLS, all modes are excited a-priori and all modes having wave-length above the laser one are detected,[16] so that a global fluctuation response should be chosen (see e.g. the chapters by S. O. Demokritov and V. E. Demidov and by G. Gubbiotti *et al.*). On the opposite limit only a spatially averaged response is accessed in inductive measurements: in the case of dots situated under the center conductor of a coplanar waveguide, the measured inductive coupling is simply proportional to an in-plane diagonal component of spatially averaged susceptibility tensor $\langle \chi_{xx}(\omega) \rangle$ (see e.g. Ref. 3 for a comparison of different excitation geometries). Finally time resolved imaging techniques provide a quantity which can be understood as the convolution of the experimental resolution with either a diagonal or non-diagonal local susceptibility depending on the excitation scheme (coplanar waveguide or current loop) and on the detection scheme (polar or longitudinal magnetization contrast). For example, the coplanar waveguide/XPEEM data of Raabe *et al.* gives access to $\chi_{xx}(\mathbf{r})$,[35] the current loop/transmission x-ray data of Stoll *et al.* and the current loop/TR-MOKE data of Perzlmaier *et al.* give access to $\chi_{zz}(\mathbf{r})$[34,40] while the coplanar waveguide/polar Kerr data of Park *et al.* give access to $\chi_{zx}(\mathbf{r})$.[33] Note finally that each measurement has a limited bandwidth, so that only part of the absorption spectrum may be accessed.

6. Theoretical Description of the Normal Modes of the Landau State

Once having reproduced some experimental results using numerical simulations, one still has to interpret the observations as simply as possible. This will be done in this section with the help of (semi)-analytical quantitative descriptions, which — though approximate — capture the physics governing each kind of mode.

6.1. *A description of the vortex mode*

In thin film elements in the flux closed state of the magnetization (for example disks or squares) the vortex which forms at the center of the element exhibits a special dynamic excitation. In the case of disks this excitation — the so-called gyrotropic vortex mode, or translational mode — has been extensively studied both experimentally and theoretically. In the case of squares in the Landau state, the general topology of this excitation remains similar to the one of the gyrotropic mode in disks, however, its description is complicated by the presence of the four domain walls. In general, for large but thin elements (e.g. with an aspect ratio D/d (Diameter/thickness) larger than 100, while $D > 1$ μm) the gyrotropic mode and the magneto-static modes can be regarded as decoupled. For smaller aspect ratios significant coupling between the translation mode of the vortex and the magneto-static modes exists which can be evidenced in the lifting of the degeneracy of some magneto-static modes (for details see Ref. 24). This behavior has been found and described in particular for disk shaped elements. In the following we will restrict ourselves to the uncoupled case for large elements.

The gyrotropic mode can be described as follows. We describe the element by three parameters: 1) the sense of circulation of the magnetization around the center C (C = +/− 1); 2) the direction of the polarization of the vortex core P (up or down) and 3) the vorticity of the element (vortex or antivortex). In the case of thin film elements in the shape of disks or squares the vorticity of the element is always fixed to a vortex. In the absence of an external bias field the vortex core is located at the center of the disk or square. After the application of an in-plane magnetic field pulse the vortex is displaced from the center. While the system relaxes towards its primary equilibrium state, the vortex gyrates around the center of the element. With element dimensions in the micrometer range and thicknesses around 10–20 nm, the frequencies of this gyrotropic mode lie in the range of several ten to several 100 MHz. The sense of rotation of the vortex at fixed vorticity depends only on the polarization of the vortex (up or down) and is dominated by the total force acting on the vortex which is given by $\mathbf{F} = \mathbf{G} \times \frac{d\mathbf{R}}{dt} - \frac{\partial W(\mathbf{R})}{\partial \mathbf{R}}$. Here, $\mathbf{R}$ is the distance of the vortex from the center, $\mathbf{G}$ is the gyrovector with a direction into or out of the plane (depending on the sign of the polarization P),and W is the total energy of the vortex. The total energy of the vortex does neither depend on the polarization nor on the chirality and thus the direction of the gyrotropic motion depends solely on the polarization of the vortex as has been shown experimentally by Choe *et al.* (squares),[8] Park and Crowell (disks)[32,33] and van Wayenberge *et al.* and Chou *et al.* (squares).[9,43] A comparison between experiment and theory has been given (for the case of disks) by Guslienko *et al.*[18]

6.2. *A description of domain-wall modes*

As mentioned in Section 2, domain-wall modes can be seen as the result of the confinement of spin waves into the potential well associated with the low values of the equilibrium in the domain-wall core. However, to be made quantitative, this approach would require a self-consistent treatment of the exchange and dipole fields such a mode would create onto itself, which becomes quite complicated when the equilibrium magnetization itself is strongly inhomogeneous. In order to circumvent this complication, we resort to a slightly different approach;[4] following early studies on domain-wall resonance,[30,38] we assume the wall is able to move as a whole without changing its in-plane internal structure $\mathbf{M}(\xi, \tau)$. This corresponds to a substitution:

$$\mathbf{M}(\xi, \tau) \leftarrow \mathbf{M}(\xi + \epsilon(\tau), \tau), \tag{5}$$

using the coordinate system (O, ξ, τ) shown in Fig. 9(d).

This quasi-static deformation of the domain-wall preserves its core structure, with the consequence that there are no extra in-plane dipole-exchange fields generated on the short scale. The fundamental domain-wall mode can therefore be understand as a Goldstone translational mode of its core. For calculating its frequency, we follow again the theory of domain-wall resonance: we shall associate a mass and a spring constant to the motion of the DW core.[30] In our case, the mass is associated with the out of plane effective field: the faster the domain-wall moves in the plane, the more its magnetization tilts out of plane, which increases accordingly the magnetic energy (dominated by its demagnetizing part). On the other hand, the spring constant is associated with the long-range dipole fields induced by the deformation of the domain-wall. The wall is indeed pinned at the dot corner (because a motion of its extremity along one edge would produce a redhibitory amount of surface magnetic charges)[21] and at the dot center (as the energy scale of the vortex is much lower that that of the domain-wall, we assume the vortex is dynamically dead at the frequency of interest). As a consequence, the wall moves by bending itself. As this bending happens on quite a large scale, the generated effective field is mostly dipolar in nature (see Fig. 9(e) for a sketch of the magnetic volume charges and demagnetizing in-plane fields generated by the bending of the DW).

To make this approach quantitative, we resort to a variational calculation.[20,46] We chose the following ansatz for the in-plane dynamic magnetization:

$$m_{\tau,n}(\xi,\tau) = m_0 \frac{1}{4\xi^2 + \delta_{DW}^2} \cos \frac{\pi n \tau}{l_{DW}}, \tag{6}$$

where δ_{DW} and $l_{DW} = \frac{w}{\sqrt{2}}$ stand respectively for the width and the length of one domain-wall and n is a non-zero integer counting the number of antinodes for the wall bending (see Fig. 9(e–g) for $n = 1 - 3$). The lorentzian dependence is chosen because it reproduces fairly well the expected transverse profile for the in-plane dynamic magnetization (see Fig. 9(e)). It is also in qualitative agreement with the measured profile published by Stoll *et al.*[40] The sinusoidal bending is chosen for simplicity. This ansatz is finally injected into a Kittel like formula (zeroth order of a variational approach):

$$f_n^{DW} = \frac{|\gamma|}{2\pi} \mu_0 M_s \sqrt{\langle G_{\tau\tau}\rangle_n \langle G_{zz}\rangle_n}, \tag{7}$$

where the $G_{\alpha\alpha}$'s are obtained by projecting the Kernel of the dipolar interaction Eq. (3) onto the ansatz Eq. (6). Note that the dynamic exchange fields do not play a significant role in determining the mass and spring constant for reasons mentioned above.

Using such a procedure for the DW described in Fig. 9 ($w = 3$ μm, $t = 30$ nm permalloy dot), one obtains a fundamental DW frequency $f_1^{DW} = 1.5$ GHz. This frequency accounts correctly for the lowest frequency peak DW_1 which is observed both in the measured and simulated spectra in Fig. 8(a). Moreover, this frequency lies well below the set of domain modes (which start at about 3 GHz) and well above the expected frequency for the vortex translational mode. Our initial assumption of a

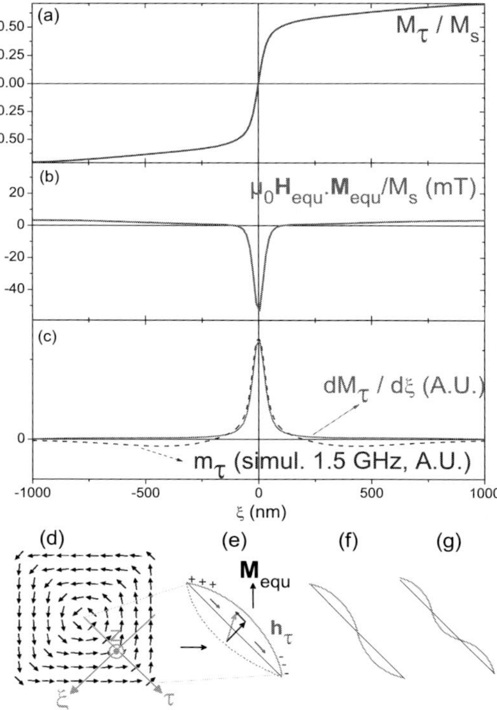

Fig. 9. Static and equilibrium properties of a domain-wall in the Landau state (same parameters as in Fig. 8). Panels (a) and (b) show respectively the transverse equilibrium magnetization and equilibrium effective field profiles extracted from the micromagnetic simulation. Panel (c) shows a comparison of the dynamical in-plane magnetization profiles deduced from Eq. (5) (continuous line) and from the 1.5 GHz simulated susceptibility map shown in Fig. 8(b) (dashed line). (d) Definition of the system of coordinates. (e)–(g) Sketch of the bending for one domain-wall. The distorted magnetization configuration is also sketched in panel (e), together with the induced magnetic charges and the corresponding in-plane dipole field h_τ. (Taken from Ref. 4.)

specific domain-wall translational mode only loosely coupled to the domains and the vortex is therefore fulfilled.

The procedure described above is applicable as long as a clear domain-wall core structure is present. As no analytical theory for the domain-walls in the Landau state is presently available, an equilibrium simulation is required for determining the width of the domain-wall core (which is presumably quite dependent on the dimensions and magnetic parameters of the dot). Quite naturally, the in-plane demagnetizing factor $\langle G_{\tau\tau} \rangle$ of

Eq. (7) is expected to decrease steadily as a function of the length of the domain-wall. This explains the size dependence of the domain-wall mode frequency measured by Park *et al.*,[33] but also the applied field dependence measured in Ref. 4. This also explains the relatively high frequency (13 GHz) simulated by Yan *et al.* for the fundamental DW mode in relatively small dots.[45] In the latter case, the domain walls were even close enough to each other so that their coupling becomes significant, thus giving rise to a set of coupled modes with different symmetries.[e] This procedure also allows one to describe higher order modes [see Figs. 9(f) and 9(g)] which appear very similar to the flexural modes measured in stripe domain systems.[44] However, these higher frequency modes may fall within the set of domain modes, so that they may be significantly renormalized by their coupling to the domains.[4]

6.3. *A description of domain modes*

For describing modes extending into the domains, one is tempted to establish a simple quantization procedure, as in the case of saturated rectangular dots (see the chapter by S. O. Demokritov and V. E. Demidov). This would require to know the behavior of the dynamic magnetization on all the edges of the domain (i.e. on the dot edges, on the vortex and on the DWs). For the physical border of the dots, we shall use the dipole exchange boundary condition of Guslienko *et al.*[21] and for the vortex core, we shall follow recent works on radial and azimuthal modes in disks.[19,24] While these works suggest a non zero amplitude at the dot boundary and/or at the vortex for small dots (such as the ones simulated in Refs. 41, 45), this translates into a vanishing of the dynamic magnetization both a the dot edges and a the dot center for relatively large dots such as that of Fig. 8. In the latter case, the wave vector at the center of each domain is required to fulfil:

$$k_x = n\frac{2\pi}{w}, \tag{8}$$

where k_x is the component of the wave-vector perpendicular to the equilibrium magnetization, n is a non-zero integer and $w/2$ is the width of the domain. Let us then plot the dispersions $f(k_{x,n}, k_y)$, where k_y is the component of the wave-vector parallel to the equilibrium magnetization and $f(\mathbf{k})$ is the dipole-exchange spin wave dispersion[25] [see Fig. 10(b)]. Such a plot provides a convenient way to classify the modes.[6] In particular,

[e] Note also that low frequency modes localized specifically at the corners appear in simulations of dots with perfect corners.[41,45]

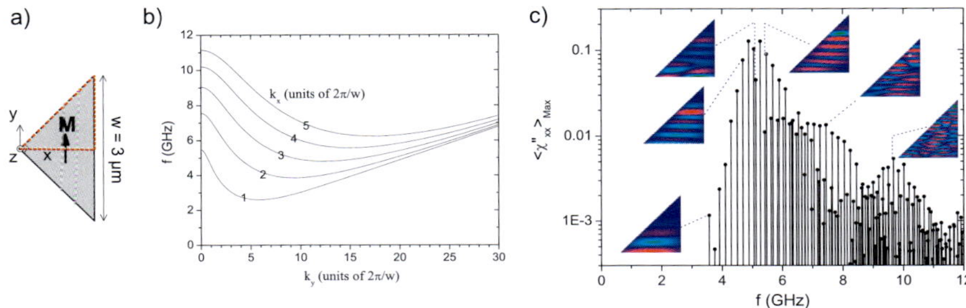

Fig. 10. Description of the domain modes for a permalloy square dot in the Landau configuration (same parameters as in Fig. 8). (a) System of coordinates used for the description of one domain. (b) Quasi-dispersions $f(k_{x,n}, k_y)$ for spin waves propagating in one domain. The strong pinning boundary condition Eq. (8) has been inserted in the dipole-exchange spin wave dispersion of Kalinikos and Slavin.[25] (c) Spectrum deduced from the diagonalization of the dynamical matrix of an "isolated" domain (i.e. a triangular dot saturated along its base). The maximum level of the spatially averaged susceptibility is reported for each mode. Susceptibility maps are also reported for a few selected modes [only the upper half of the triangle is shown, see the dotted line in panel (a)]. (Some data taken from Ref. 4.)

it shows how the anisotropy of the quantization process derives from the anisotropy of the dipole-exchange dispersion itself. One recognizes indeed a transition between high frequency modes having nodal lines along $\mathbf{M}_{equ}$ (i.e. having wave-vector mostly perpendicular to the magnetization, corresponding to spin waves of the magnetostatic surface wave type)[10,25] and the low frequency normal modes having nodal lines perpendicular to $\mathbf{M}_{equ}$ (i.e. having wave-vector mostly parallel to the magnetization, corresponding to spin waves of magnetostatic backward volume wave type).[10,25] Figure 10(b) suggests in particular that the lowest frequency mode corresponds to the minimum of the $n = 1$ dispersion lying at $f = 3$ GHz and $k_y = 6\frac{2\pi}{w}$. This is confirmed by inspecting the 2.9 GHz susceptibility map in Fig. 8(b).

In trying to go one step further, i.e. predict quantitatively the position and intensity of absorption peaks such as those of Fig. 8(a), one would have to derive a quantization procedure accounting for the presence of the domain-walls. Unfortunately, there is currently no description of the specific behavior of the dynamic magnetization in the domain-wall region. From the inspection of the equilibrium configuration (see Fig. 9), it is clear that the spin waves are strongly accelerated in the domain-wall core because of the very low value of the equilibrium effective field in this region (see e.g. the WKB calculation by Bayer *et al.*[5] for short wavelength spin waves propagating through a 180° Bloch domain-wall). However, this picture is complicated by the fact that neighboring domains are also connected via long-range dipole fields which depend on the full 2D distribution of the magnetization (and therefore of the symmetry of the mode considered).[f] Using a simplified ansatz for the dynamic magnetization in one domain and accounting variationally for the (symmetry dependent) long range dipole fields coupling neighboring domains and domain-walls,[4] we could reproduce the frequency of the main peaks observed in Fig. 8(b) (the result of this calculation are displayed as arrows on top of the plot). Such a rough estimate is however far from describing the full modal structure of the domains, which is presumably quite complicated. By drawing an analogy with billiard systems (which constitute the classical mechanics counterpart of our systems),[4] one can indeed convince oneself that a confined spin wave problem displays a simple modal structure only in very special cases (such as the fully elliptical and rectangular geometries). For generic geometries, the problem is "non-integrable",[4] which has two important practical consequences: First one cannot derive any simple quantization procedure describing the full domain mode spectrum [with respect to this, Fig. 10(b) should only be seen as a skeleton over which the normal modes are built but there is no systematic rule for building them]. Second the spectrum consists of many modes looking quite similar to each other. As a consequence, a given excitation may couple to many modes having nearly the same frequencies. This is illustrated in Fig. 10(c), which shows the result of the diagonalization of the dynamical matrix for a fictitious "isolated" domain (i.e. an isorectangle triangle saturated along its hypothenuse). Many similar modes appear to combine together to form a broad feature around 5 GHz. We expect such an irregular spectrum also in the case of the domain modes of the full Landau structure: the complicated reflection-refraction rule occurring when a spin wave meets a domain-wall play qualitatively the same role as the total reflection occurring at the edges of the saturated triangle. The practical consequences of this lack of regularity are quite dependent on the size of the dot. For relatively small dots,[6] the frequency spacing between adjacent modes can remain larger than the intrinsic linewidth so that these modes could be resolved experimentally. For relatively large dots, on the other hand, the observed peaks are expected to be composite ones, which

[f] For illustration purpose, the domain spectrum of the Landau structure could probably be viewed as an intermediate between the spectrum of an "isolated" domain (case in which the discontinuity imposed by DWs is exaggerated) and the spectrum of a circular dot in the vortex state (case in which the discontinuity imposed by DWs is suppressed).

is confirmed by inspecting the susceptibility maps in Fig. 1(b) where one recognizes some change of nodal structure across e.g. the D_1^0 peak or the D_2 one.

7. Summary and Outlook

The experimental and theoretical works reported in this chapter combine together to form a global picture for the normal modes of the Landau configuration. These normal modes are naturally deeply influenced by the equilibrium structure itself: the micromagnetic equilibrium "singularities" (vortex and domain-walls) exhibit their own dynamics while the shape of the domains affects deeply their quantized spin wave spectrum.

This picture could certainly be refined in the future. This will probably require detailed comparisons of experiments, simulations and analytical results for different dot sides and thickness. By analogy with the recent discoveries for circular disks, couplings between the vortex mode, the domain-wall modes and the domain modes should be observable. Finally, this normal mode -linear- picture should be extended to include nonlinear effects if one wishes to address large amplitude dynamics such as vortex switching,[43] domain-wall trapping,[28] field-driven magnetization reversal[31] or thermally activated switching.[13]

References

1. http://llgmicro.home.mindspring.com.
2. Object oriented micro magnetic framework, www.nist.gov.
3. Bailleul, M., Höllinger, R. and Fermon, C. (2006), Microwave spectrum of square Permalloy dots: Quasi-saturated state, *Phys. Rev. B* **73**, p. 104424.
4. Bailleul, M., Höllinger, R., Perzlmaier, K. and Fermon, C. (2007), Microwave spectrum of square Permalloy dots: Multi-domain state, *Phys. Rev. B* **76**, p. 224401.
5. Bayer, C., Schultheiss, H., Hillebrands, B. and Stamps, R. L. (2005), Phase Shift of Spin Waves Traveling Through a 180° Bloch-Domain Wall, *IEEE Trans. Mag.* **41**, pp. 3094–3096.
6. Bolte, M., Meier, G. and Bayer, C. (2006), Spin-wave eigenmodes of Landau domain patterns, *Phys. Rev. B* **73**, p. 052406.
7. Buess, M., Raabe, J., Perzlmaier, K., Back, C. H. and Quitmann, C. (2006), Interaction of magnetostatic excitations with 90° domain walls in micrometer-sized permalloy squares, *Phys. Rev. B* **74**, p. 100404(R).
8. Choe, S. B., Acremann, Y., Scholl, A., Bauer, A., Doran, A., Stöhr, J. and Padmore, H. A. (2004), Vortex Core-Driven Magnetization Dynamics, *Science* **304**, pp. 420–422.

9. Chou, K. W., Puzic, A., Stoll, H., Dolgos, D., Schtz, G., Van Waeyenberge, B., Vansteenkiste, A., Tyliszczak, T., Woltersdorf, G. and Back, C. H. (2007), Direct observation of the vortex core magnetization and its dynamics, *Appl. Phys. Lett.* **90**, p. 202505.
10. Damon, R. W. and Eshbach, J. R. (1961), Magnetostatic modes of a ferromagnet slab, *J. Phys. Chem. Solids* **19**, pp. 308–320.
11. Demidov, V. E., Demokritov, S. O., Hillebrands, B., Laufenberg, M. and Freitas, P. (2004), Radiation of spin waves by a single micrometer-sized magnetic element, *Appl. Phys. Lett.* **85**, pp. 2866–2868.
12. DeSimone, A., Kohn, R. V., Müller, S. and Otto, F. (2002), A reduced theory for thin-film micromagnetics, *Comm. on Pure and Appl. Math.*, pp. 1408–1460.
13. E, W., Ren, W. and Vanden-Eijndena, E. (2003), Energy landscape and thermally activated switching of submicron-sized ferromagnetic elements, *J. Appl. Phys.* **93**, pp. 2275–2282.
14. Grimsditch, M., Giovannini, L., Montoncello, F., Nizzoli, F., Leaf, G. K. and Kaper, H. G. (2004a), Magnetic normal modes in ferromagnetic nanoparticles: A dynamical matrix approach, *Phys. Rev. B* **70**, p. 054409.
15. Grimsditch, M., Leaf, G. K., Kaper, H. G., Karpeev, D. A. and Camley, R. E. (2004b), Normal modes of spin excitations in magnetic nanoparticles, *Phys. Rev. B* **69**, p. 174428.
16. Gubbiotti, G., Carlotti, G., Okuno, T., Grimsditch, M., Giovannini, L., Montoncello, F. and Nizzoli, F. (2005), Spin dynamics in thin nanometric elliptical Permalloy dots: A Brillouin light scattering investigation as a function of dot eccentricity, *Phys. Rev. B* **72**, p. 184419.
17. Guslienko, K. Y., Demokritov, S. O., Hillebrands, B. and Slavin, A. N. (2002), Effective dipolar boundary conditions for dynamic magnetization in thin magnetic stripes, *Phys. Rev. B* **66**, p. 132402.
18. Guslienko, K. Y., Han, X. F., Keavney, D. J., Divan, R. and Bader, S. D. (2006), Magnetic vortex core dynamics in cylindrical ferromagnetic dots, *Phys. Rev. Lett.* **96**, p. 067205.
19. Guslienko, K. Y., Scholz, W., Chantrell, R. W. and Novosad, V. (2005), Vortex-state oscillations in soft magnetic cylindrical dots, *Phys. Rev. B* **71**, p. 144407.
20. Guslienko, K. Y. and Slavin, A. N. (2000), Spin-wave excitations in cylindrical magnetic dot arrays with in-plane magnetization, *J. Magn. Magn. Mater.* **215-216**, pp. 576–578.
21. Guslienko, K. Y. and Slavin, A. N. (2005), Boundary conditions for magnetization in magnetic nanoelements, *Phys. Rev. B* **72**, p. 014463.
22. Hillebrands, B., Ounadjela, K. and Thiaville, A. E. (2001–2006), *Spin Dynamics in Confined Magnetic Structures*, Vol. 1–3 (Springer).
23. Hubert, A. and Schäfer, R. (1998), *Magnetic Domains* (Springer).
24. Ivanov, B. A. and Zaspel, C. E. (2005), High frequency modes in vortex-state nanomagnets, *Phys. Rev. Lett.* **94**, p. 027205.

25. Kalinikos, B. A. and Slavin, A. N. (1986), Theory of dipole-exchange spin wave spectrum for ferromagnetic films with mixed exchange boundary conditions, *J. Phys. C: Solid State Phys.* **19**, pp. 7013–7033.
26. Kaya, A. and Bain, A. J. (2006), High frequency susceptibility of closure domain structures calculated using micromagnetic modeling, *J. Appl. Phys.* **99**, p. 08B708.
27. Knox, R. S. and Gold, A. (1964), *Symmetry in the solid state* (Benjamin).
28. Krasyuk, A., Wegelin, F., Nepijko, S. A., Elmers, H. J., Schonhense, G., Bolte, M. and Schneider, C. M. (2005), Self-trapping of magnetic oscillation modes in Landau flux-closure structures, *Phys. Rev. Lett.* **95**, p. 207201.
29. Landau, L. and Lifshitz, E. (1935), On the theory of the dispersion of magnetic permeability in ferromagnetic bodies, *in Collected papers of L. D. Landau, Gordon and Breach* (*Ter Haar Ed., 1965*), pp. 101–114.
30. Malozemoff, A. P. and Slonczewski, J. C. (1979), *Magnetic Domain Walls in Bubble Materials* (Academic).
31. Neudert, A., McCord, J., Schfer, R. and Schultz, L. (2005), Subnanosecond vortex transformation in ferromagnetic film elements observed by stroboscopic wide-field Kerr microscopy, *J. Appl. Phys.* **97**, p. 10E701.
32. Park, J. P. and Crowell, P. A. (2005), Interactions of spin waves with a magnetic vortex, *Phys. Rev. Lett.* **95**, p. 167201.
33. Park, J. P., Eames, P., Engebretson, D. M., Berezovsky, J. and Crowell, P. A. (2003), Imaging of spin dynamics in closure domains and vortex structures, *Phys. Rev. B* **67**, p. 020403(R).
34. Perzlmaier, K., Buess, M., Back, C. H., Demidov, V. E., Hillebrands, B. and Demokritov, S. O. (2005), Spin-wave eigenmodes of permalloy squares with a closure domain structure, *Phys. Rev. Lett.* **94**, p. 057202.
35. Raabe, J., Quitmann, C., Back, C. H., Nolting, F., Johnson, S. and Buehler, C. (2005), Quantitative analysis of magnetic excitations in Landau flux-closure structures using synchrotron-radiation microscopy, *Phys. Rev. Lett.* **94**, p. 217204.
36. Ramesh, M. and Wigen, P. E. (1988), Ferromagnetodynamics of parallel stripe domains — domain walls system, *J. Magn. Magn. Mater.* **74**, pp. 123–133.
37. Riedel, H. and Seeger, A. (1971), Micromagnetic treatment of Nel walls, *Phys. Stat. Sol.* **(b)46**, pp. 377–384.
38. Slonczewski, J. C. (1972), Dynamics of magnetic domain walls, *Intern. J. Magnetism* **2**, pp. 85–97.
39. Smit, J. and Wijn, H. P. J. (1959), *Ferrites* (Philips Technical Library).
40. Stoll, H., Puzic, A., van Waeyenberge, B., Fisher, P., Raabe, J., Buess, M., Haug, T., Höllinger, R., Back, C., Weiss, D. and Denbeaux, G. (2004), High-resolution imaging of fast magnetization dynamics in magnetic nanostructures, *Appl. Phys. Lett.* **84**, pp. 3328–3330.
41. Vaast-Paci, C. and Leylekian, L. (2001), Numerical simulations of isolated particles susceptibilities: effects of shape and size, *J. Magn. Magn. Mater.* **237**, pp. 342–361.

42. van den Berg, H. A. M. (1986), Self-consistent domain theory in soft-ferromagnetic media. II. Basic structures in thin film objects, *J. Appl. Phys.* **57**, pp. 2168–2173.
43. van Wayenberge, B., Puzic, A., Stoll, H., Chou, K. W., Tyliszczak, T., Hertel, R., Fähnle, M., Brückl, H., Rott, K., Reiss, G., Neudecker, I., Weiss, D., Back, C. H. and Schütz, G. (2006), Magnetic vortex core reversal by excitation with short bursts of an alternating field, *Science* **444**, pp. 461–464.
44. Vukadinovic, N., Serraj, A., Le Gall, H. and Ben Youssef, J. (1998), Dynamic susceptibility of parallel stripe domins with flexing domain walls, *Phys. Rev. B* **58**, pp. 385–393.
45. Yan, M., Leaf, G., Kaper, H., Camley, R. and Grimsditch, M. (2006), Spin-wave modes in a cobalt square vortex: Micromagnetic simulations, *Phys. Rev. B* **73**, p. 014425.
46. Zivieri, R. and Stamps, R. L. (2006), Theory of spin wave modes in tangentially magnetized thin cylindrical dots: A variational approach, *Phys. Rev. B* **73**, p. 144422.

Chapter 5

SPIN WAVES IN THE INHOMOGENEOUS INTERNAL FIELD OF NANOSTRUCTURED RINGS

Dirk Grundler

Physik Department, Technische Universität München, Garching, Germany

Fabian Giesen

Max-Born-Institut, Berlin, Germany

Jan Podbielski

Institut für Angewandte Physik, Universität Hamburg, Hamburg, Germany

1. Introduction

The understanding of the dynamics and the mode spectrum of ferromagnetic micro- and nanostructures has attracted great attention for both basic and applied research. In recent years the evolution of lithography and characterization techniques has enabled the fabrication of well-tailored magnets and therefore the investigation of fundamental aspects of spin dynamics.[41,42] Initial studies focussed on quantized waves in wires and disks where the geometrical edges impose boundary conditions.[45,56] Soon, it was realized that the inhomogeneous internal magnetic field leads to additional localization phenomena[44] or forms a barrier for tunneling of spin waves.[20] Additionally, the inhomogeneous field of a domain wall has been predicted to shift the phase of a propagating spin wave.[9,39] At the same time the computational power has increased in such a manner that magnetodynamic simulations can be performed for individual and small arrays of micro- and nanomagnets on a desktop computer. This helps to interpret experimental data, identify characteristic contributions to spin dynamics, i.e., from the exchange energy and dipolar interactions, and predict novel behavior. In applied research and magnetic storage technology, ultrafast switching of a magnet with small energy dissipation is key. Switching is triggered e.g. by a transient magnetic field pulse which drives a large angle motion of the magnetization. If the switching is dominated by precession (as opposed to relaxation) it can be as fast as 165 ps.[68] The uttermost switching speed is obtained when the magnetic moments rotate coherently represented by one giant vector (macrospin) on a ballistic trajectory. Pioneering work that has promoted the basic understanding of precessional switching to date is based on very different magnetic systems and experiments. Amongst them there are plain ferromagnetic films excited by accelerated electron bunches,[3,4,74] spin valve devices consisting of magnetic multilayers[68,69] and individual ferromagnetic microstructures.[5,24,40] Microwave assisted switching of a crystalline magnetic nanocluster has been observed[73] opening a further pathway for switching with sub-Stoner-Wohlfarth reversal fields. We expect both the shape of the magnet and the inhomogeneous internal field profile to be crucial for the achievement of reliable ultrafast switching. For this, a detailed understanding of spin precession eigenmodes in inhomogeneous magnetic fields is of great importance.

We describe recent experiments on microwave spectroscopy of nanorings. Starting from the discovery of ferromagnetic resonance (FMR) in 1946[30] and stimulated by early theoretical work on this phenomenon[48,49] FMR experiments have been found to be powerful to determine the magnetic properties of magnetized materials.[22,37] A conventional FMR setup uses a resonant cavity of a fixed microwave frequency where the magnetic sample is inserted.[76] Sweeping an external magnetic field generates the resonance condition for power absorption at the spin precession eigenfrequency. The cavity and its resonance frequency thus rule the magnetization state that is monitored. Often the saturated state is addressed. If one is, however, interested in the dynamics of the different characteristic magnetization states of a micromagnet and of an inhomogeneous spin configuration at remanence, a technique would be indispensable where instead of the magnetic field the frequency is swept. We have set up a *broadband* microwave spectrometer that allows us to measure the dynamics of an arbitrary magnetic configuration.[12,28,63] The apparatus follows an all-electrical excitation and detection scheme. It measures the dynamic response excited by a high frequency magnetic field. In our experiments, the time-varying field is a sinusoidal (continuous) wave $\vec{h}_{\rm rf}$. A coplanar waveguide (CPW) optimized for the GHz frequency regime guides the excitation field to the magnetic sample. The response is inductively picked up by the same CPW and detected in the frequency domain by a vector network analyzer (VNA).[75] We will call this technique VNA-FMR in the following. The approach is complementary to the pulsed inductive microwave magnetometer or PIMM[70] that is operated in the time domain and where a transient pulse $\vec{h}_{\text{pulse}}$ excites the magnetic sample. By using a large pulse amplitude it is possible to enlarge the precession angle beyond the switching limit of the magnet. We will use the VNA-FMR technique since it is more sensitive and allows us to study the small precession-angle regime with a high signal-to-noise ratio. Our broadband spectrometer is also complementary to techniques like Brillouin light scattering (BLS) in the frequency domain and the magneto-optical Kerr effect in the time domain,[1,19,23,40,52] which both have been used recently as an imaging technique with submicron resolution.

Over the last years, the static magnetization configurations of ferromagnetic rings have attracted great interest.[51,55,66] A magnetic ring exhibit both a stray field free (flux-closure) configuration (vortex state) and a polarized configuration (onion state). In the latter state, the two ring arms are magnetized in parallel and incorporate two domain walls. The ring is hence an intriguing system to investigate homogeneous and inhomogeneous configurations in the same device at remanence. The material permalloy ($Ni_{80}Fe_{20}$) is advantageous as it does not exhibit an intrinsic magnetic anisotropy. The ring's properties are then controlled by only a small set of geometrical parameters which are thickness, width and diameter. The static magnetization states are often found to be stable against unavoidable imperfections introduced by the film growth and patterning process. The ring geometry is thus very interesting to investigate fundamental aspects of spin dynamics on the nanometer scale. Recently rings together with disks have been considered to be building blocks for magnonics.[14,39] Here, spin wave interference effects are a central issue.[63]

We will review VNA-FMR experiments on single-layered rings prepared from permalloy.[28,29,32,62,63,77,78] By this, we outline the current understanding of spin wave discretization and localization, and we make clear how spin excitations are tailored via geometrical boundaries and the inhomogeneous internal field H_{int}.

2. Fundamentals of Spin Waves: From a Plain Film to Microscopic Rings

In this chapter we present an introduction to the basic physics of spin waves in microstructures. We discuss how the lateral patterning modifies the wave-vector dispersion of spin excitations. For the sake of simplicity we will start studying the case of a thin ferromagnetic film. Then we proceed to a mesoscopic wire before, finally, we address a ring-like structure. In this way, we introduce successively the fundamental aspects which govern the spin excitations in microscopic rings.

Let us start from a ferromagnetic thin film where (i) the magnetization M is saturated due to a large external field H and (ii) the static internal field H_{int} is homogeneous. In this case a low-energy collective spin excitation can exhibit a wave vector $k = 0$ where all spins precess coherently around H. This excitation is called uniform spin precession and might be detected by an FMR experiment. In case of a BLS experiment, one is able

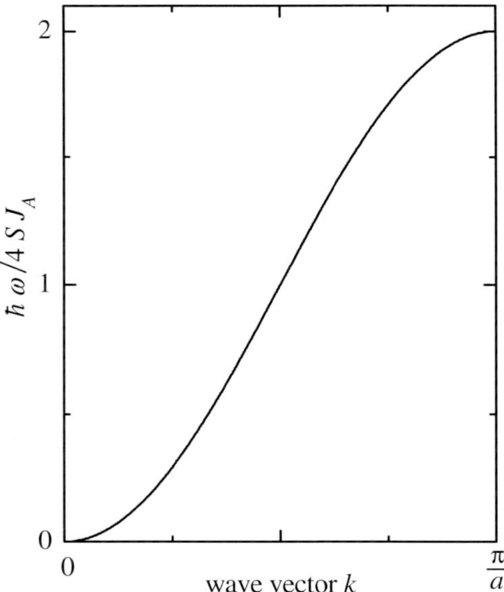

Fig. 1. Wave-vector dispersion $\omega(k)$ of a spin wave in a ferromagnet exhibiting a homogeneous internal field, i.e., the magnet is saturated and geometrical boundaries do not play a role. The frequency is given in units of S (spin) and J_{A} (exchange integral). For clarity a one-dimensional ferromagnet is assumed.

to transfer momentum to the spin system. Thereby the excitation of spin waves with $k \neq 0$ becomes possible. Such excitations, called magnons, propagate within the magnetic film and exhibit a characteristic dispersion $\omega(\vec{k})$. A typical dispersion curve that is often displayed in solid state textbooks is shown in Fig. 1. For clarity, we assume a one-dimensional ferromagnet with nearest-neighbor spin-spin interaction. The eigenfrequency of the spin wave increases monotonously with increasing wave vector k. Here it is important that only exchange contributions are considered. Approaching the Brillouin zone boundary at $k = \frac{2\pi}{\lambda} = \frac{\pi}{a}$, where a is the lattice constant, the wavelength λ becomes so short that neighboring spins are misaligned. Exchange interaction dominates the high eigenfrequency. For spin waves in microstructures it is important to consider also contributions which arise from both static and dynamic demagnetization fields, i.e., from dipolar interaction. In this scenario the eigenfrequencies do not depend only on the absolute value of the wave vector $|\vec{k}|$ but also on the angle α enclosed by $\vec{k}$ and the magnetization $\vec{M}$. The solid lines in Fig. 2(a) are calculated following Ref. 47 and include both dipolar and exchange interactions. The depicted dispersion curves are valid for a sample where the magnetization $\vec{M}$ lies in the plane of the film and is collinear with the external magnetic field $\vec{H}$ [cf. Fig. 2(b)]. Spin waves with $\vec{k} = (k_x, k_y, k_z)$ propagate without restrictions only in the x, y plane. The top and bottom film surfaces impose a boundary condition on the wave vector k_z. In z direction a standing spin wave is formed and k_z becomes discrete according to $k_z = \frac{p\pi}{t}$. Here, t is the film thickness and $p = 0, 1, 2, \ldots$. We will discuss the lowest mode with $p = 0$ only. Considering unpinned spins at the surfaces the eigenmode is uniform in z direction. For the in-plane directions we now assume $k^2 = k_x^2 + k_y^2$. We distinguish two configurations in Fig. 2(a) depending on the angle α between $\vec{k}$ and $\vec{M}$. The angle turns out to be key for the spin wave dispersions $f(k) = \frac{\omega(k)}{2\pi}$: in case of $\alpha = \frac{\pi}{2}$, i.e. $\vec{k} \perp \vec{M}$, the eigenfrequency f is rising monotonously with k. These spin waves are called Damon-Eshbach modes (DE modes). In contrast, when $\alpha = 0$, i.e. $\vec{k} \parallel \vec{M}$, f shows a negative slope with k for small wave vectors. This is a so-called backward volume magnetostatic wave (BVMSW). The term "backward" originates from the negative group velocity $v_{\text{gr}} = \frac{\partial \omega}{\partial k}$. The characteristics of both types of spin waves at small wave vector k (i.e. for long wavelengths) are dominated by the dipolar interactions which are of long range. This wave-vector regime is thus termed "dipolar regime" in the literature. Here the contribution of the exchange energy is proportional to Jk^2 (J is

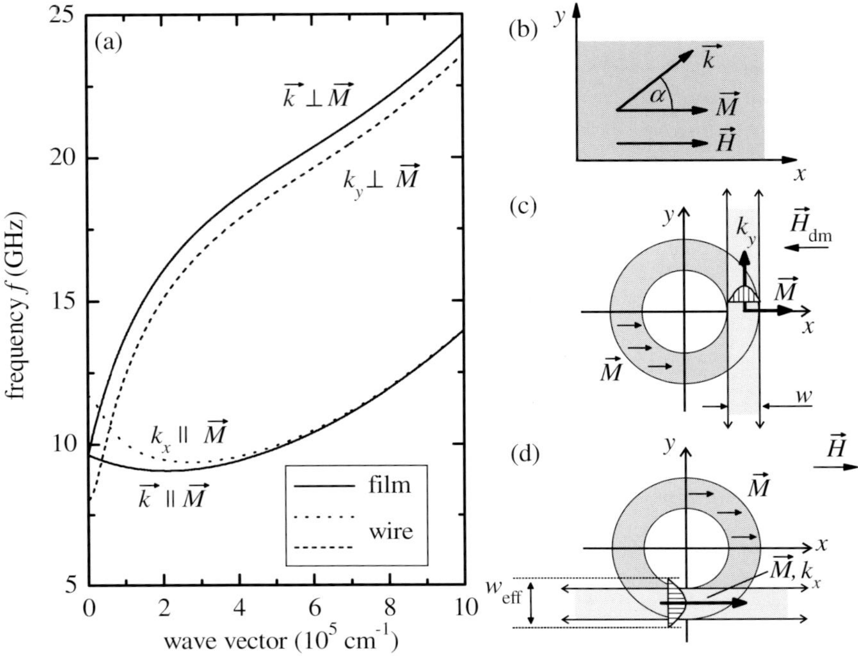

Fig. 2. (a) Wave-vector dispersions at $\mu_0 H = 100$ mT for an infinite film for two orientations: $\vec{k} \parallel \vec{M}$ (bottom solid trace) and $\vec{k} \perp \vec{M}$ (upper solid trace). The dashed (dotted) trace reflects a wire with $w = 700$ nm which is magnetized transversely (longitudinally). (b)-(d) Sketches of the relative orientations of $\vec{H}$, $\vec{M}$ and $\vec{k}$ in a plain film (b), a transversely (c) and longitudinally (d) magnetized wire (light gray area). The graphs in (c) and (d) suggest that infinite wires might be regarded as limiting cases to describe the spin dynamics within a saturated ring (dark gray area).

the exchange constant and can be calculated from the exchange integral J_A used in Fig. 1). Thus only for large values of k the exchange interaction takes over and leads to a quadratic increase of the corresponding eigenfrequencies. For high values of k both types of spin waves are therefore termed to be "exchange dominated". The two different dispersion curves displayed as solid traces in Fig. 2(a) reflect the upper and lower band edges of the spin wave band. Dispersion curves at angles $0 < \alpha < \frac{\pi}{2}$ fall between these band edges. Close to the Brillouin zone boundary the dispersions approach the characteristics of Fig. 1.

Patterning a ferromagnetic film on the micrometer scale and into a magnetic wire leads to significant changes of the dynamic properties, as the magnet's geometrical boundaries create in-plane demagnetizing fields. These fields compete with the external field H (sometimes called the Zeeman field) and the effective exchange field. In Fig. 2(c) we show a transversely magnetized wire (light gray area) and assume a DE mode propagating along y direction. Two aspects are important: (i) The static demagnetizing field $H_{\rm dm}$ reduces the internal field $|H_{\rm int}| = |H| - |H_{\rm dm}|$. This results in a shift of DE modes to lower frequencies [cf. dashed trace in Fig. 2(a)]. The shift does not depend on k. (ii) The finite width w of the wire leads to a wave-vector quantization of k_x. For small k, this quantization induces an additional shift to even smaller f. To calculate the full dispersion curve $f\,(k_y)$ (dashed trace) in Fig. 2(a) we follow Ref. 34 and assume a cosine spin wave profile along the x direction. This is depicted in Fig. 2(c). We take $k_x = 2\pi/w$, i.e., the length $\frac{\lambda}{2}$ fits into the width. The dashed trace is below the dispersion of the plain film. Based on Fig. 2(d) we now address a wire magnetized along its long axis. In this scenario there is no static demagnetizing field and $|H_{\rm int}| \simeq |H|$. The characteristics of the BVMSW propagating along the wire are changed however by dynamical dipolar effects. They lead to a modification of the dispersion as shown by the dotted trace in Fig. 2(a). The eigenfrequencies are increased if compared to the corresponding dispersion of the plain film. This is valid, however, only in the dipolar regime, i.e., at small wave-vectors k. For the longitudinally magnetized wire one finds that the wave vector component k_y is quantized. The exact profile of this standing spin wave strongly depends on the pinning conditions at the edges of the wire. Guslienko *et al.* found that dipolar effects cause a pinning which depends on the geometrical dimensions like width and thickness.[35] Spins which are located close to a wire's edge interact more intensively with the dynamical demagnetizing field than spins in the center of the wire. To quantify the pinning strength the effective width $w_{\rm eff} > w$ was introduced as illustrated in Fig. 2(d). The quantization condition then reads $k_y = \pi/w_{\rm eff}$.

The dispersion curves displayed in Fig. 2 already provide a basis to understand qualitatively the spin dynamics observed for microscopic magnetic rings. Let us assume a ring where $\vec{M}$ is saturated along the x direction [cf. Fig. 2(c)]. A spin wave excited in the head and tail exhibits the character of a DE mode, i.e., propagation occurs with k_y perpendicular to $\vec{M}$. In the side arms however [cf. Fig. 2(d)] the spin wave properties reflect a BVMSW as k_x is pointing along $\vec{M}$. The character of a spin wave which travels around a saturated ring thus varies from DE-type to BVMSW-type back and forth depending on the azimuthal position. Excitation frequencies fall between the dashed and dotted dispersion traces. The ring's symmetry and peculiarities of the internal field H_{int} thus give rise to novel spin wave physics. After discussing our experimental techniques in the following section we will present a more detailed theoretical approach and experimental data on the spin wave physics of rings in the remaining part of this chapter.

3. Experimental Techniques

In this section we provide information on both measurement and preparation techniques. First, the broadband spectrometer for the magnetization dynamics measurements is described. Second, patterning of thin-film magnets by means of electron beam lithography and lift-off processing is discussed. Magnetic and atomic force microscopy (MFM and AFM, respectively) are used in particular for the characterization of our samples.

3.1. *Broadband spectroscopy*

For the excitation and detection of magnetization dynamics of individual micro- and nanostructures or arrays of them we use a broadband GHz spectrometer operating at room temperature. It is depicted in Figs. 3(a) and 3(b). The CPW is optimized to exhibit low losses and a microwave impedance of $Z_0 = 50$ Ohm.[17] This ensures high field amplitude h_{rf} at the sample's location and the efficient detection of the inductive response. Magnets are integrated either directly on the GaAs substrate or on the central conductor of the CPW [Fig. 3(c)]. In both cases the magnetic devices are defined by electron beam lithography and lift-off processing. This will be presented in detail in Section 3.2. The CPW is fabricated from a metallic sandwich which consists of Cr, Ag and Au. Typically it is 160 nm thick. The gold serves as a caplayer to prevent the silver from oxidization. The thickness is chosen such that the Ohmic losses of the waveguide are small. The rms roughness of the CPW typically varies between 2.5 and 5 nm as measured by AFM. We use commercial microwave probes attached to micropositioners to connect the CPW to the VNA. The pitch of probes and contact pads is 250 μm. A microscope is needed to find the optimum position for the probes and thus minimize reflections [Fig. 3(b)]. Two orthogonal electromagnets, each with a maximum field of 100 mT, generate a field $\vec{H}$ in the plane of the magnetic sample. This field is used to define the magnetic history. To ensure optimum mechanical stability, the whole setup

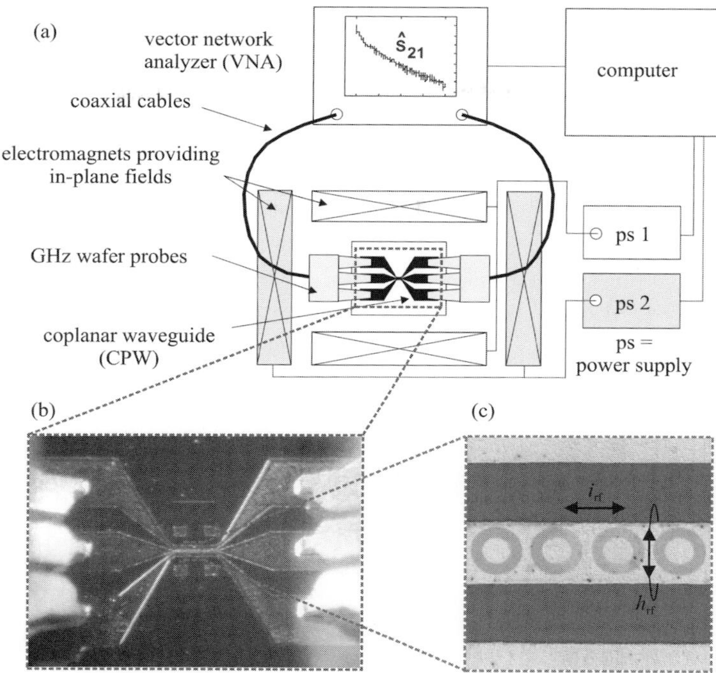

Fig. 3. (a) Block diagram of the broadband GHz spectrometer using a vector network analyzer. The VNA measures the complex scattering parameters (transmission: $\hat{S}_{21}$; reflection: $\hat{S}_{11}$) of a coplanar waveguide containing patterned ferromagnets. The CPW resides within two electromagnets generating an in-plane vector field. (b) GHz probes (yellow color) connect the CPW (light gray) which is tapered to a width of a few microns in its center. (c) AFM picture of a single row of 2.2 μm wide permalloy rings on top of the 2.9 μm wide central conductor of a CPW. The ground leads are seen at the top and bottom. The CPW (yellow) is prepared on semi-insulating GaAs (red).

is mounted on an optical table with an active vibration damping system. It is remotely controlled by a personal computer.

In the following, we discuss details of the broadband spin wave spectroscopy using the VNA. In the experiment a sinusoidal high frequency current $i_{\rm rf}$ is passed through the CPW and produces the rf magnetic field $\vec{h}_{\rm rf}$. If $\vec{h}_{\rm rf}$ exerts a torque $\vec{\tau} = \vec{M} \times \vec{h}_{\rm rf}$ on the magnetization $\vec{M} = \vec{M}_0 + \vec{m}(\omega, t)$ this gives rise to a time dependent magnetization $\vec{m}(\omega, t)$ that is orthogonal to $\vec{M}_0$. In the following, we will label the static magnetization by $\vec{M}_0$ and the dynamic component by $\vec{m}(\omega, t)$. Note that in our devices both quantities depend on the spatial coordinates, since we deal with inhomogeneous magnetization configurations, an inhomogeneous internal field and, as we will see, a finite wave-vector $\vec{k}$ of the spin excitations. These aspects are all different if compared to the early work on FMR.[30,48,49,76]

It is instructive to review the spatial profile of the excitation field. In the CPW we expect quasi-TEM modes to be present. As a consequence the field component in the propagation direction is negligible: $h_x \ll (h_y, h_z)$. Let us first assume a uniform current density over the CPW's central conductor with width w_C. The VNA-current induced field then lies almost perfectly in the plane of the magnets and perpendicular to the direction of the waveguide. In this case we obtain the magnetic field from $h_{\rm rf} = i_{\rm rf}/2w_C$.[70] However, at GHz frequencies the current density is expected to be inhomogeneous and exhibit extremal values at the edges due to the self-inductance of the metallic conductors.[43] Using free commercial software[71] we model the current distribution at 20 GHz[59] and determine the profile of $\vec{h}_{\rm rf}$ according to Biot and Savart's law [Fig. 4]. The in-plane component h_y is

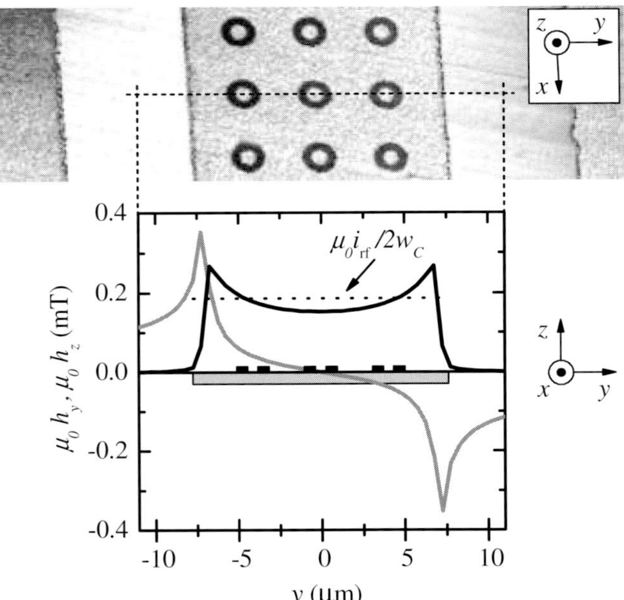

Fig. 4. Upper panel: Scanning force micrograph of a CPW containing an array of rings on the central conductor. Lower panel: We compare the field components h_y calculated for a very low-frequency (broken line) and an rf current $i_{\rm rf}$ at 20 GHz (full black line). The gray curve depicts the component h_z.

found to be almost constant over the central conductor where the magnets are placed. We do not expect the enhancement of the in-plane field at the edges to play an important role as long as we choose the excitation field strength such that the linear regime of spin dynamics is addressed. The out-of-plane component h_z shows a high absolute value at the edges only outside the periodic arrays of nanomagnets. We therefore disregard this inhomogeneity when analyzing the experimental data. To substantiate this assumption we have performed simulations considering both the homogeneous and realistic excitation field profile. We have not obtained a difference in the calculated spin-wave absorption spectra.

The dynamics stimulated by $\vec{h}_{\text{rf}}$

$$\vec{m}(\omega, t) = \hat{\chi}(\omega)\vec{h}_{\text{rf}}(\omega, t)$$

induce a voltage V_{ind} in the waveguide. In the case of inhomogeneous magnetization patterns or rf excitation fields $\vec{m}$ depends also on the spatial coordinate $\vec{r}$ and on the wave-vector $\vec{k}$. According to Ref. 70 V_{ind} is calculated from reciprocity according to

$$V_{\text{ind}} = \frac{\mu_0 N}{2} \int_{\text{magnet}} \vec{h}_{\text{rf}} \cdot \frac{\mathrm{d}\vec{m}(\vec{r}, t)}{\mathrm{d}t} \, \mathrm{d}\vec{r}. \tag{1}$$

Here, N is the number of magnets on the CPW and $\hat{\chi}$ the tensor of the complex magnetic susceptibility. The integration is performed over one of the magnets. Note that V_{ind} might be measured in reflection or in transmission geometry. Equation (1) shows that due to the distinct rf field profile present in a CPW only the transverse component of the dynamic magnetization, $\vec{m}_t = (m_y, m_z)$, is excited and detected. If magnets exhibit a small aspect ratio "thickness/width" m_y exceeds m_z by the inverse scaling factor, so the detection signal is dominated by m_y. When the dynamic magnetization is at resonance the phase shift between the exciting rf current i_{rf} and the induced current $i_{\text{ind}} = Z_0^{-1} V_{\text{ind}}$ is π. As a consequence absorption peaks occur at resonant spin wave eigenmodes of the magnets.

To extract the small inductive signal given by Eq. (1) from the complex electrodynamic response of the CPW including microwave probes and coaxial cables two experimental methods are applied: on the one hand we perform a thorough calibration of the whole setup and on the other hand we apply a difference detection scheme. The calibration is a two-step process. First a reference sample containing lithographically well-defined thin-film microwave devices like a through, an open and a matched load is used to calibrate the setup according to the so called thru-reflection-match (TRM) method.[2] From this calibration we obtain parameters that remodel the dynamic response of the cables and connectors including the microwave probes. Second we replace the reference substrate by the CPW containing the nanomagnets and pursue the TRM calibration in such a manner that we consider the CPW as a through. In particular we apply a large in-plane magnetic field $\vec{H}$ perpendicular to the CPW's central conductor. In this case $\vec{M}$ and $\vec{h}_{\rm rf}$ are set to be parallel. This prevents us from exciting spin wave dynamics and $V_{\rm ind}$ is at minimum (see discussion below). After refining the characteristic TRM parameter set by this means the measurement technique is optimized to a degree that spin wave absorption features are seen directly on the display of the VNA. In our case the relative signal strength is about 10^{-3} to 10^{-2} dB.

We use a difference detection scheme to further improve the signal-to-noise and standing wave ratio.[15,16,29,70] Prior to taking data under resonant conditions we acquire a reference spectrum with $\vec{M}_0$ being collinear with $\vec{h}_{\rm rf}$ to a very large extent. In this case, excitation and detection of magnetization dynamics are vanishingly small because (i) the torque on the static magnetization is tiny, i.e., $\vec{\tau} = \vec{M}_0 \times \vec{h}_{\rm rf} \approx 0$, and (ii) the dynamic magnetization, if present, is oriented such that the induced voltage $V_{\rm ind}$ is very small. These arguments hold also for the second step of the TRM calibration scheme. Unsaturated edge regions in the magnets might exhibit $\vec{M}_0 \parallel$ CPW$\perp \vec{h}_{\rm rf}$, however the fraction of spins in such regions is assumed to be small. The voltage induced by an out-of-plane precessional component m_z is expected to be tiny also. One finds $\vec{h}_{\rm rf} \cdot {\rm d}\vec{m}/{\rm d}t \approx i\omega h_z m_z \approx 0$. This holds because $\vec{h}_{\rm rf}$ is mainly in plane of the CPW and $h_z \approx 0$ at the position of the magnets (cf. Fig. 4 and Eq. (1)). As a consequence saturation in a direction perpendicular to the CPW provides a reference spectrum where absorption features due to spin waves are at minimum or even absent. This reference spectrum is then subtracted from the dynamic response measured in the desired magnetization configuration $\vec{M}$. The difference detection scheme is advantageous also because in long-term measurements we have observed that the electrical parameters drift and that the calibration degrades. Variations in the ambient conditions (room temperature, humidity) or magnetic field driven displacements of probes and cables might be responsible for the observed drifts.

3.2. *Preparation of ferromagnetic nanostructures*

To optimize the inductive coupling we integrate ferromagnetic samples to coplanar waveguides[28] or transmission lines[52] directly by means of electron beam lithography and lift-off processing.[33,38,57] The devices are prepared on semi-insulating (001) GaAs substrates since they provide a surface with a very small roughness in the Ångström range and their dielectric constant is in particular very well specified. This allows us to design optimized CPWs with low losses. We have used nanomagnets fabricated both directly on top of a GaAs substrate and on top of a 160 nm thick metallic conductor of a CPW to investigate the impact of surface roughness on the dynamic properties. In the latter case, the increased surface roughness of the CPW which is in the hundreds of Ångström range changes the morphology of the investigated nanomagnets. The physics on which we report here do not depend on whether the magnets were fabricated on GaAs and then monolithically covered by the CPW's central conductor or vice versa.

The typical preparation steps for magnets on a CPW are as follows. First, we fabricate a coplanar waveguide using a chrome mask, UV lithography and lift-off processing of a 160 nm thick metallic multilayer. The outer contact pads are as wide as 100 μm to accommodate the microwave probes having a pitch of 250 μm. We taper the central conductor to a width w_C of a few microns in the central part of the waveguide. We note that fabrication of the whole CPW in one and the same exposure and lift-off processing step is crucial. When using a mix-and-match technique leading to overlapping metallic conductors within the CPW we observe a high loss, which increases with microwave frequency, and numerous internal reflections. Second, after optical inspection of the CPW we use predefined markers to expose the nanomagnets on the tapered central conductor using electron beam lithography. The resist consisting of PMMA is optimized to exhibit an undercut profile after development. This facilitates the subsequent lift-off processing. Third, we evaporate $Ni_{80}Fe_{20}$ (permalloy). The base pressure in the high vacuum chamber is 10^{-7} mbar. Typically we deposit a permalloy thickness smaller than 35 nm which ensures that the microscopic magnetic moments are oriented in the plane of the substrate. A thin cover layer of Au (5 nm) or resist is added to some samples to prevent the permalloy's surface from oxidizing. It has already been reported that storing permalloy nanodevices in ambient atmosphere might change quasistatic magnetic properties and switching fields.[54] Such variations would be detrimental in the long-term course of time-consuming broadband spin wave spectroscopy on nanomagnets. A thorough inspection using optical microscopy and atomic force microscopy follows the lift-off processing to verify the thickness and geometry of the magnets. These data are essential to perform decisive micromagnetic simulations since magnetic properties and dynamics are governed by the exact shape.

4. Modeling: Micromagnetic Simulations versus Semi-Analytical Approach

We both perform micromagnetic simulations (MS) and develop a semi-analytical (SA) modeling to analyze experimental eigenmode spectra and to interpret the data.[27] Combining the two theoretical approaches is instructive to clarify the microscopic physics underlying confinement, localization and interference phenomena of spin waves. In the following we outline the basics of our theoretical modeling. To substantiate the discussion we present results which help to illustrate the spin wave dynamics in rings which we investigate in the following chapters.

4.1. *Micromagnetic simulations*

For the simulations the magnets are subdivided into a mesh with a discrete number of cells where the side lengths are chosen to be no larger than the exchange length. This is about 5 nm for permalloy. The small cell size is essential in order not to average over the spatial variation of both the magnetization $\vec{M}(x, y, z)$ and effective magnetic field $\vec{H}_{\text{eff}}$. The equilibrium state of a magnet is obtained by integrating numerically the Landau-Lifshitz (LL) equation[53]

$$\frac{\mathrm{d}\vec{M}^i}{\mathrm{d}t} = -|\gamma|\mu_0\vec{M}^i \times \vec{H}^i_{\text{eff}} + \frac{|\gamma|\alpha\mu_0}{M_S}\vec{M}^i \times (\vec{M}^i \times \vec{H}^i_{\text{eff}}). \qquad (2)$$

This equation is assumed to be valid for the macrospin $\vec{M}^i$ of each individual cell i with effective field $\vec{H}^i_{\text{eff}}$. γ represents the spectroscopic splitting factor and α the phenomenological damping parameter. The cells are interconnected via exchange and dipolar interactions. The latter competes with the exchange interaction and possible magnetic anisotropies and leads to the long-range order of microscopic moments. The effective field $\vec{H}_{\text{eff}} = (1/\mu_0)\partial E/\partial\vec{M}$ considers the different energy terms in the equation of motion (Eq. (2)). Here, E is the total energy density containing demagnetization, exchange, anisotropy and Zeeman energy. The equilibrium domain configuration of mesoscopic permalloy magnets is ruled by dipole-dipole interaction and reflects the magnet's shape in a characteristic manner (cf. Fig. 1 in Ref. 31). The dispersion $\omega(\vec{k}) = 2\pi f(\vec{k})$ of long wavelength spin waves are governed by dynamic dipolar coupling of magnetic moments. The spin wave phenomena discussed here fall mainly in the category of such dipolar modes. Following Section 2 we distinguish in

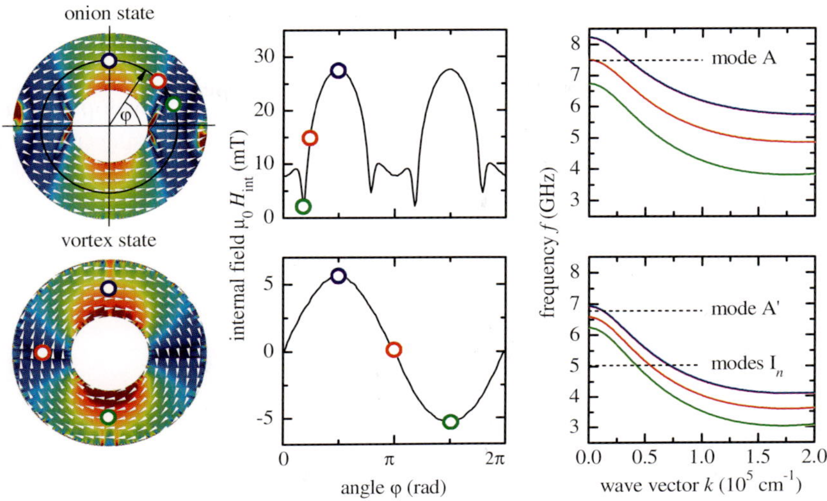

Fig. 5. Upper row from left to right: Static magnetization configuration of the onion state at $\mu_0 H = 30$ mT. In the simulation performed by OOMMF we take a width $w = 700$ nm, an outer diameter $2R_o = 2.2\ \mu$m, a thickness $t = 25.5$ nm and characteristic material parameters for permalloy. Small white arrows indicate the local orientation of the magnetization $\vec{M}$. Blue (red) color represents a small (large) *absolute* value of the internal field $\vec{H}_{int}$. The central graph depicts the azimuthal dependence of the internal field H_{int}. The right graph contains spin-wave dispersions $f(k) = \omega(k)/(2\pi)$ for the three different field values H_{int} highlighted in the central graph. Lower panel from left to right: Data set simulated in the vortex state where an in-plane field $\mu_0 H = 5$ mT is considered. The vortex state is hence polarized. In the left graph the color coding of H_{int} is different from the top row as far as absolute values are concerned. Note that in the central graphs the sign of H_{int} reflects the relative orientation between $\vec{H}_{int}$ and the local $\vec{M}$. The field $\vec{H}$ is applied in the horizontal direction, pointing to the right. The spin wave modes labeled by A, A′ and I_n will be discussed in Section 5.

particular between the backward volume magnetostatic wave (BVMSW) where $\vec{k}$ is *collinear* to $\vec{M}$ and the Damon-Eshbach-like (DE) spin wave where $\vec{k}$ is *perpendicular* to $\vec{M}$. In the course of our investigations we have used different software and codes[10,21,67] to perform such kind of numerical calculations. Magnetic parameters like $|\gamma| = 28$ GHz/T and the saturation magnetization M_S measured on reference films are taken as input parameters. For the exchange constant we use $J = 13 \cdot 10^{-12}$ J/m. Based on these data we simulate the equilibrium magnetic configuration at a given magnetic field. For details on micromagnetic simulations see Ref. 21.

In the left column of Fig. 5 we show static magnetization configurations of the onion and the vortex state. In the center column we depict the

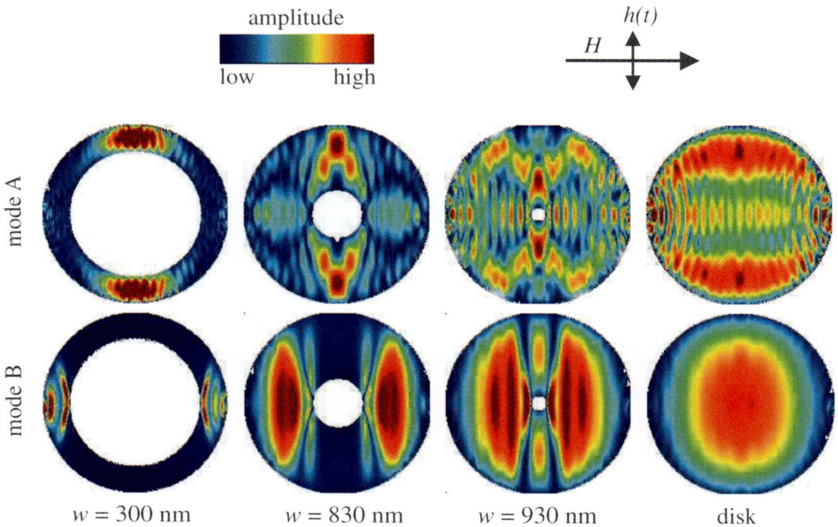

Fig. 6. Spin wave mode patterns calculated by micromagnetic simulations. From left to right the width w increases. Rings are symmetric, i.e., the hole is in the center, and in the onion state at $\mu_0 H = 90$ mT. The pattern of a microdisk is shown on the right. The outer diameter is taken to be 2 μm. The eigenfrequencies are larger in the top row (modes of type A) if compared to the bottom row (modes of type B). For details see text.

simulated internal fields $H_{\rm int}$. The sign of the internal field is defined by the magnetization orientation: $H_{\rm int}$ is negative (positive) for an antiparallel (parallel) alignment of $\vec{M}$ and $\vec{H}_{\rm int}$. $H_{\rm int}(\vec{r})$ will be essential to understand the spin dynamics in the investigated nanomagnets. Equilibrium configurations as shown in Fig. 5 are secondly used as the initial state for the simulation of the dynamics. To evaluate frequencies up to 20 GHz the LL equation is integrated with time steps of the order of 25 ps and for a time duration of several ns. The time evolution of each cell $\vec{m}(x_i, y_i, z_i, t)$ is recorded after a delta-pulse type of excitation $\vec{h}_{\rm pulse}(t) = h_0\delta(t)\vec{e}_y$ ($\vec{e}$ denotes a unit vector). Then we perform a fast Fourier transformation and extract the spectral amplitude at a given frequency for each cell (Fourier transform imaging).[13] This method allows us to visualize eigenmodes. In Fig. 6 spin wave mode patterns for different rings in the onion state and for a disk are shown. Here we depict the main modes A and B which will be discussed in Sections 5.1 and 5.3. For the top (bottom) row the corresponding eigenfrequencies are large (small). Since we use a delta-type $h_{\rm pulse}$ the calculated excitation spectrum must be weighted by the frequency f

as is the case for the experimentally measured induced voltage $V_{\rm ind}$. Then a direct comparison between simulated and measured spectra is possible.[25] The damping constant α is typically 0.5 (0.01) for simulations of the static (dynamic) properties. Spatial profiles of the static internal field are separately simulated using the free OOMMF code[21] but otherwise identical simulation parameters. Comparisons between OOMMF, the LLG solver and MicroMagus reveal similar profiles.

Micromagnetic simulations are helpful to interpret an individual spectrum and, as depicted in Fig. 6, to visualize the segments of a magnet where modes exhibit the largest spin wave amplitude. Varying the geometrical parameters by a small amount needs a further full set of simulations that are time consuming, and the dependence of spin-wave eigenfrequencies on the simulation parameters is not always fully clear. Here a semi-analytical approach is often instructive. It can help to provide a detailed understanding if the dependence of spin wave eigenfrequencies on a characteristic parameter is addressed. The semi-analytical approach is found to be in particular powerful to clarify the evolution of spin wave modes as a function of a geometrical parameter like the ring width w. We will derive spin-wave dispersion curves $f(k) = \omega(k)/(2\pi)$ and use them to clarify in detail the localization process in rings and to model the spin wave interference effects. We emphasize that we take the semi-analytical model as a complementary means with respect to our micromagnetic simulations where an individual device is modeled. We interconnect both approaches when we start the semi-analytical calculations from internal fields $\vec{H}_{\rm int}$ extracted from micromagnetic simulations.

4.2. *Semi-analytical approach: Overview*

We follow earlier semi-analytical approaches which have been developed to describe modes in an inhomogeneously magnetized stripe,[6,8,46,61] in a spin wave well[44] and in a rectangular micromagnet.[7,34] To find a spin wave spectrum in a semi-analytical approach one solves the equation

$$\frac{\mathrm{d}\vec{M}}{\mathrm{d}t} = -|\gamma|\mu_0 \vec{M} \times \vec{H}_{\rm eff}, \tag{3}$$

which, compared to Eq. (2), omits the damping term. The effective magnetic field is given by

$$\vec{H}_{\rm eff} = \vec{H} + \frac{2J}{\mu_0 M_S^2}\nabla^2 \vec{M} + \int_V \hat{G}(\vec{r},\vec{r}\,')\vec{M}(\vec{r},\vec{r}\,')\mathrm{d}\vec{r}\,'. \tag{4}$$

The first term $\vec{H}$ represents the external magnetic field and the second term reflects the exchange field. The magnetization, defined by $\vec{M} = \vec{M}_0 + \vec{m}(\vec{k},t)$, contains static and dynamic contributions. The third term incorporates the tensorial Green's function $\hat{G}(\vec{r},\vec{r}\,')$ which relates both the dynamic magnetization to the dynamic demagnetizing field and the static magnetization to the static demagnetizing field. We use the term *effective* field H_{eff} for the field that is inserted in the LL equation and includes in particular *all* static and dynamic magnetic field contributions. The *internal* field H_{int} denotes the sum of the static fields, e.g., external, anisotropy and demagnetizing fields, that are present in a ferromagnetic element, excluding the exchange field. In the absence of crystalline anisotropy the internal field reduces to the *demagnetization* field H_{dm}. One obtains the following integro-differential equation:

$$\begin{aligned}&\left(-B\omega_M\nabla^2 + \omega_{H_{\mathrm{int}}}\right)\hat{I}\vec{m}(\rho,\varphi) + i\omega\hat{T}\vec{m}(\rho,\varphi) \\ &\quad -\omega_M\int \hat{G}_{\rho,\varphi}(\vec{r},\vec{r}\,')\vec{m}(\rho',\varphi')\rho'\mathrm{d}\rho'\mathrm{d}\varphi' = 0\end{aligned} \tag{5}$$

with $B = \frac{2J}{\mu_0 M_S^2}$, $\omega_M = |\gamma|\mu_0 M_S$, and $\omega_{H_{\mathrm{int}}} = |\gamma|\mu_0 H_{\mathrm{int}}$. The matrices $\hat{I}$ and $\hat{T}$ are given by

$$\hat{I} = \begin{pmatrix} 1 & 0 \\ 0 & 1 \end{pmatrix}, \quad \hat{T} = \begin{pmatrix} 0 & -1 \\ 1 & 0 \end{pmatrix}. \tag{6}$$

Certainly the set of equations can be solved numerically. However the computational effort would be not much different from performing full-scale micromagnetic simulations. More insight is gained if a semi-analytical solution is provided. For specific geometries even a fully analytical solution might be possible. In the following we work out an appropriate solution that allows us to understand microscopic details and parameter dependencies of spin wave phenomena in nanorings.

4.3. *Semi-analytical approach: Formalism for rings*

To adapt the semi-analytical approach to rings and find a solution of the LL equation we start from a plane wave with a spatially varying wave-vector

$$\vec{m}(\vec{k},t) = \vec{m}_{mn}(\vec{r})\exp[i(\vec{k}_{mn}(\vec{r})\cdot\vec{r} - \omega_{mn}t)]\,. \tag{7}$$

The values $m = 0,1,2,\ldots$ and $n = 0,1,2,\ldots$ will be used later to quantify the number of nodal lines in orthogonal spatial directions. Since the frequency of an eigenmode ω_{mn} is constant throughout the magnet, the spin-wave wave-vector varies spatially to accommodate the changes in the inhomogeneous internal field profile $\vec{H}_{\text{int}}(\vec{r})$:

$$\omega_{mn} = \omega_{mn}\left(\vec{H}_{\text{int}}(\vec{r}), \vec{k}_{mn}(\vec{r})\right) = \text{const.} \tag{8}$$

The main challenge is to find the most appropriate dispersion relation $\omega(k)$ as the exact one is not known a priori. Once $\omega(k)$ is known for different magnetic fields, the spatial dependence of the wave-vector can be evaluated by the following means: for each position x, y within the magnet we consider the local value of the internal field $H_{\text{int}}(x, y)$ and plot the corresponding spin-wave dispersion. We consider only in-plane directions since we deal with thin permalloy nanomagnets in the following, i.e., $\vec{r} = x \cdot \vec{e}_x + y \cdot \vec{e}_y$ [cf. Section 2]. The intersection of the 'local' dispersion curve with the considered eigenfrequency gives the corresponding local value of the wave-vector $k(x, y)$.[44] Then we map $k(x, y)$ for the whole nanomagnet. Note that this procedure is mathematically analogous to the WKB approximation of a quantum mechanical wave function in a potential landscape. If the nanoring's aspect ratios obey $t/w \ll 1$ and $t/R \ll 1$ (thickness divided by width and radius, respectively) the dynamical dipole-dipole matrix elements in Eqs. (4) and (5) can be neglected.[7,34] In this limit the dispersion of an infinite film might hold if one starts from wave-vectors reflecting the specific symmetry of the magnet. For rectangular elements wave-vectors have been derived early, and we leave them out here.[7,34] For rings we recently considered:[27]

$$k_{m\rho} = \frac{(m+1)\pi}{w_{\text{eff}}} \quad \text{and} \quad k_{n\varphi} = \frac{(n+1)\pi}{\xi}, \tag{9}$$

where $\xi = \bar{R} \cdot \Phi_0$ is the characteristic arc length in azimuthal direction and w_{eff} the effective width of the ring. The parameter Φ_0 will be used in Section 5.2 and denotes an angular segment of the ring. In Sections 5.1 and 5.4 it is reasonable to take $\bar{R} = R_{\text{o}} - w/2$. Note that in Eq. (9) the parameter *effective* width is used and not the *geometrical* width. It has been argued in Refs. 35, 36 that due to pinning phenomena the (effective) width entering in Eq. (9) might be different from the geometrical one [cf. Section 2]. Using the wave-vectors of Eq. (9) we derive the following dispersion relation:

$$\begin{aligned} \omega_{mn}^2 = {} & (\omega_{mn}^H + B\omega_M \kappa_{mn}^2) \cdot \\ & \left(\omega_{mn}^H + B\omega_M \kappa_{mn}^2 + \omega_M F_{mn}(\kappa_{mn} t)\right). \end{aligned} \tag{10}$$

$\kappa_{mn}^2 = k_{m\rho}^2 + k_{n\varphi}^2$ denotes the total in-plane wave-vector. For the semi-analytical analysis, to start from a realistic profile of the inhomogeneous internal field in the 2D plane of a thin magnet we take the field $H_{\mathrm{int}}(\rho, \varphi)$ which we obtain by using micromagnetic simulations [see chapter 4.1]. The field is weighted with the mode profile:[34]

$$\omega_{mn}^H = |\gamma|\mu_0 \langle H_{\mathrm{int}} \rangle_{m_{mn}} = \frac{\int H_{\mathrm{int}} m_{mn}^2 \mathrm{d}^2 r}{\int m_{mn}^2 \mathrm{d}^2 r} \,. \tag{11}$$

Assuming factorization in orthogonal directions the normalized mode profile exhibits the form:

$$m_{mn}(\rho, \varphi) = \mu(k_{m\rho}, \rho) \cdot \eta(k_{n\varphi}, \varphi) \,. \tag{12}$$

In particular, the functions μ and η are expressed by cosine and sine functions for even and odd indices m, n, respectively.[34] Based on this, one calculates the dispersion and checks the consistency. Considering mode profiles of Eq. (12), ω_{mn}^H acquires the form $\omega_{mn}^H = |\gamma|\mu_0 H - |\gamma|\mu_0 \langle H_{\mathrm{dm}} \rangle_{m_{mn}}$, with

$$\langle H_{\mathrm{dm}} \rangle_{m_{mn}} = \frac{4|\gamma|\mu_0}{\xi \left(R_{\mathrm{o}}^2/2 - R_{\mathrm{i}}^2/2 \right)} \cdot \int_{R_{\mathrm{i}}}^{R_{\mathrm{o}}} \int_{(\pi-\Phi_0)/2}^{(\pi+\Phi_0)/2} H_{\mathrm{dm}}(\varphi) m_{mn}^2 (\rho - \bar{R}, \varphi) \mathrm{d}\rho \, \rho \, \mathrm{d}\varphi \,. \tag{13}$$

Here, R_{i} and R_{o} are the inner and outer radius, respectively, and $\bar{R}$ is the radius at half the ring width. The matrix element $F(\kappa_{mn} t)$ is given by[34,47]

$$F_{mn}(\kappa_{mn} t) = 1 - P(\kappa_{mn} t) \left(\frac{k_{n,\varphi}^2}{\kappa_{mn}^2} \right) + P(\kappa_{mn} t) \cdot (1 - P(\kappa_{mn} t)) \left(\frac{\omega_M}{\omega_{mn}^H + B \omega_M \kappa_{mn}^2} \right) \left(\frac{k_{m,\rho}^2}{\kappa_{mn}^2} \right) \tag{14}$$

with

$$P(\kappa_{mn} t) = 1 - \frac{1 - \exp(-\kappa_{mn} t)}{\kappa_{mn} t} \,. \tag{15}$$

The remaining challenge is to (i) guess the correct mode profile, (ii) perform the calculations and (iii) check the consistency. Once consistency is reached the semi-analytical approach is powerful to explain microscopic details of spin wave localization in an inhomogeneous magnetic field. For illustration, the right column of Fig. 5 depicts appropriate dispersion relations $\omega(k) = 2\pi f(k)$ for BVMSWs in rings in the two characteristic magnetization states, i.e., onion (top) and vortex (bottom) state. When discussing the corresponding spin excitations we will in particular distinguish between so-called localized modes existing in narrow rings and extended modes which occur when the rings become wide (cf. Fig. 6). In the framework of the semi-analytical results and dispersion curves shown in the right column of Fig. 5 the term localization is interpreted such that the spin wave eigenfrequency, if depicted as a horizontal line, does not intersect all the dispersion curves $f(k, H_{\mathrm{int}})$ generated by the inhomogeneous internal field H_{int}. The angular segment Φ_0 where the spin wave mode exists is smaller than 2π. The special formalism to describe azimuthal modes I_n which intersect all dispersion curves and extend around the whole ring will be outlined in further detail in Section 5.4 when we discuss the experimental data.

5. Spin Wave Spectroscopy on Symmetric Rings

We have studied spin excitations in different microdisks and nanopatterned rings. We have varied the diameter and the ring width. Here we stay with micromagnets having an outer diameter $2R_{\mathrm{o}} \approx 2.1$ μm. We report on several arrays, each containing between 15 and several hundreds of rings. The spin wave phenomena that we discuss reflect the behavior of the individual device. Thus we do not address magnetostatic and magnetodynamic coupling of devices. The exact number of magnets on the CPW determines the signal-to-noise ratio (SNR) but not the nature of the spin wave excitations. For completeness, we note that we have been able to measure spin excitations of a single microdisk of diameter 2.1 μm integrated to a CPW with a width $w_C = 2.3$ μm. Though the SNR is small in this experiment (roughly 2:1) this demonstrates the high sensitivity of the optimized VNA setup. In the following we address the linear regime of spin dynamics. For this, we keep the output power P of the VNA at ≤ -10 dBm.

We first focus on rings with a small width $w = 300$ nm and discuss the quasistatic magnetic configurations. We have monitored them by measuring the stray field with a magnetic force microscope. Figure 7(a) depicts the stray field in the onion state if an in-plane field of -75 mT is applied. In Fig. 7(b) we observe the two-step reversal which is typical for narrow rings having a moderate thickness t.[55,65,66] Analyzing the MFM signal in detail [Fig. 7(c)], we find $H_1^{\mathrm{sw}} \approx 14$ mT and $H_2^{\mathrm{sw}} = 40$ mT. Note that for H_1^{sw}, i.e. the transition from onion to vortex state, we observe a distribution of switching field ranging from about 9 to 15 mT depending on the individual ring. This already demonstrates that the rings prepared on the CPW are

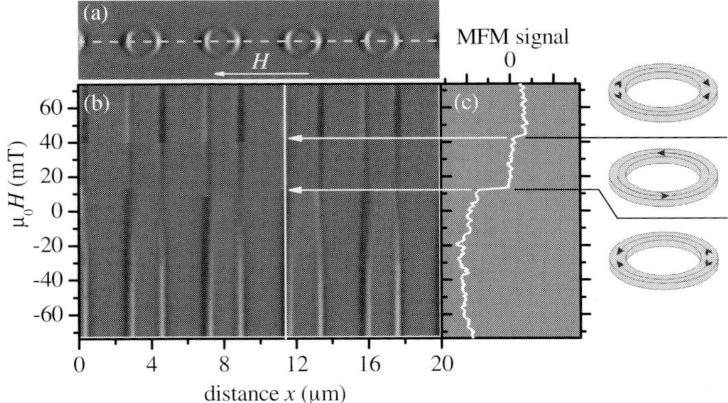

Fig. 7. (a) Magnetic force microscopy performed on rings at room temperature. White and black colors represent stray-field lines originating from local north and south poles, respectively. The field of $\mu_0 H = -75$ mT is applied in horizontal direction. (b) The graph summarizes MFM scans along the broken line depicted in (a) performed at successively increased H starting from -75 mT. These values are given on the left axis. (c) Stray field measured at the vertical line in (b). This curve reflects the magnet's hysteresis curve for increasing field H. The two-step reversal with $H_1^{\rm sw}$ [bottom white arrow in (b)] and $H_2^{\rm sw}$ [top white arrow in (b)] is clearly seen. For $H_1^{\rm sw} < H < H_2^{\rm sw}$ the ring is in the vortex state and, as expected, the stray field, i.e., the MFM signal, is nearly zero. In the outermost column on the right quasistatic magnetization configurations are sketched. Black arrows indicate the orientation of M in the upper and lower rings halves (ring arms).

not ideal and might incorporate microscopic defects or nanoindentations due to the lift-off processing of polycrystalline material. It has already been reported that $H_1^{\rm sw}$ depends on defects: in the reversal process, the domain walls of the onion state, that are initially present in the head and tail, are released from their opposing positions and move around the ring. They annihilate in order to form the domain-wall-free vortex state. Defects are able to pin a domain wall and by this means might vary the exact field strength for annihilation.[50,65] Switching to the reversed onion state at $H_2^{\rm sw}$ is however dominated by domain nucleation. This process is not much affected by microscopic details of pinning centers. Consistently the distribution of $H_2^{\rm sw}$ is narrow in Fig. 7(b).

5.1. *Narrow rings in the onion state: Localized spin waves*

In this section we present spin wave spectra measured at $\mu_0 H = 90$ mT on an array of 750 narrow rings of width $w = 300$ nm. The outer diameter

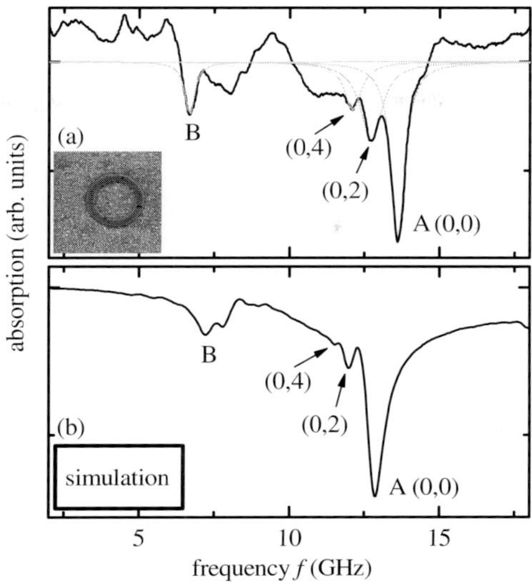

Fig. 8. (a) Measured and (b) simulated spectrum of narrow rings at $\mu_0 H = 90$ mT. The width of the permalloy rings is 300 nm (see scanning electron micrograph in the inset). Note that the simulation is performed for an individual ring at a temperature $T = 0$. Experimental data are obtained on an array at room temperature. The numbers in brackets denote the number of nodal lines as defined in Section 5.2.

$2R_\mathrm{o}$ and thickness t are 2.1 μm and 15 nm, respectively. Due to the small thickness spins are preferentially in the plane of the substrate. The magnetic field $\vec{H}$ is applied in the plane of the rings and along the CPW. Due to $H \gg H_2^{\mathrm{sw}}$ each ring is in the onion state, i.e., both ring halves are aligned with the applied field and domain walls are present in the head and tail. The torque between the magnetization $\vec{M}$ and $\vec{h}_{\mathrm{rf}}$ excites the spin waves. Figure 8(a) depicts the absorption spectrum. We summarize two important experimental findings: In the onion state of rings, (i) two main modes exist, one at high frequency (mode A) and one at low frequency (mode B). (ii) In between there are satellite modes of small intensity. The spectrum taken from the micromagnetic simulations [Fig. 8(b)] confirm the experimental observations. They are also consistent with results obtained on Co rings.[58] We note that in a further ring array we resolve up to four satellite modes. Most likely a better homogeneity of the nominally identical rings and a smaller damping allow us to observe a larger number of satellite modes.

In the following we will discuss how local details of the internal field H_{int}, the finite wave-vector k and the localization of spin waves are essential to understand the spin excitations of Fig. 8(a). For the discussion it is instructive to revisit Eq. (10) and decompose the total wave-vector κ of a spin wave in a ring into $\kappa^2 = k_\rho^2 + k_\varphi^2$. In radial direction labeled by the index ρ a quantization condition is anticipated for k_ρ due to the geometric ring borders. It is reasonable that this condition is similar to the one of a longitudinally magnetized wire[34,35,56] discussed in Section 2. It leads to discrete values $k_\rho = k_{m\rho}$. We adopt the notation from Ref. 60 and, if necessary, index the spin waves by the integer number of nodal lines m and n along the radial $\vec{e}_\rho$ and azimuthal $\vec{e}_\varphi$ directions, respectively. Important for the discussion on mode A is the azimuthal component k_φ which is parallel to the ring's magnetization. Since $\vec{k}_\varphi \| \vec{M}$ this configuration represents a BVMSW. In the following we will use the semi-analytical approach to highlight how the internal field profile which varies characteristically as a function of azimuthal angle φ gives rise to the localization of long-wavelength BVMSWs, i.e., spin waves with negative group velocity. Here it will be important that local maxima of H_{int} are found for the ring segments where spins are parallel to, both, the edges and $\vec{H}$. Local minima exist for the head and tail where microscopic magnetic moments point perpendicular to the edges. Here spin waves with a positive group velocity, i.e., DE-like spin wave modes, become localized.

5.2. *Narrow rings: Comparison between experimental and semi-analytical results*

Following the semi-analytical ansatz outlined in Section 4.3 we calculate mode A for a ring with $w = 300$ nm and $t = 15$ nm at $\mu_0 H = 90$ mT. These geometrical parameters lead to an effective width $w_{\text{eff}} = 338$ nm.[35] The internal field H_{int} at $\bar{R}$ along the azimuthal direction $\vec{e}_\varphi$ is taken from micromagnetic simulations and presented in Fig. 9(a). Here we mark the positions for which we calculate the dispersion curves shown in 9(b). Considering the experimentally observed eigenfrequency $f_{\text{exp}} = 13.6$ GHz of mode A we find that it does not exhibit an intersection with all of the displayed dispersion curves. Only inside the ring segment between the vertical grey lines (positions x_l and x_r) intersections exist. Here, the wave-vector k_φ exhibits a real value. Outside, the wave-vector becomes imaginary, and as a consequence the spin wave is damped exponentially. Note that k_φ becomes *smaller* for *decreasing* H_{int}. Figures 9(a) and 9(b) based on semi-analytical calculations are in particular powerful to illustrate the microscopic origin underlying the localization of mode A in narrow rings. The spin wave with

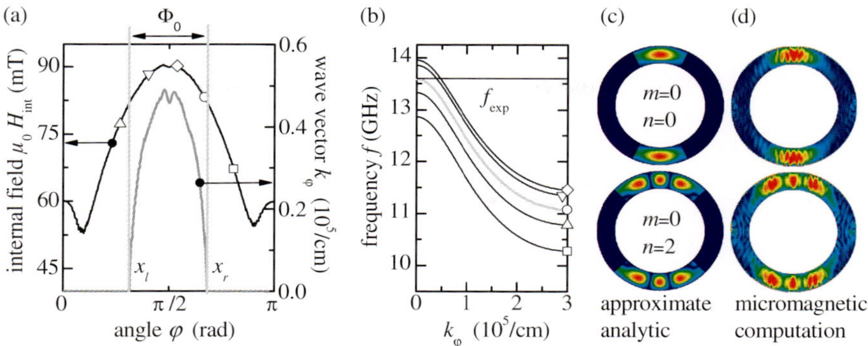

Fig. 9. (a) Profile of the internal field $H_{\rm int}(\varphi)$ at $\bar{R} = R_{\rm o} - w/2$ extracted from micromagnetic simulations (w = 300 nm) performed at $\mu_0 H$ = 90 mT. Symbols label the values $H_{\rm int}(\varphi)$ for which the corresponding dispersion curves $\omega(k_\varphi) = 2\pi f(k_\varphi)$ in (b) have been calculated semi-analytically. Based on the dispersion curves shown in (b) we conclude that the mode at eigenfrequency $f_{\rm exp}$ exhibits real wave-vectors k_φ only within the ring's segment Φ_0 depicted in (a). Comparison between mode patterns for $(m, n) = (0, 0)$ (top) and $(m, n) = (0, 2)$ (bottom) that are (c) calculated semi-analytically and (d) simulated. The field $\mu_0 H$ = 90 mT is applied in horizontal direction in (c) and (d).

eigenfrequency f_A exists only near the local maximum of the internal field $H_{\rm int}$. $H_{\rm int}$ is inhomogeneous since the onion state and H break the radial symmetry of the rings. In Fig. 9(a) the points x_l and x_r are taken to be turning points for the localized BVMSW. The symmetry of the onion state with two rings halves magnetized in parallel is such that the scenario discussed for the ring segment near $\varphi = \pi/2$ in Fig. 9 holds also for the segment near $\varphi = 3\pi/2$. Hence, the localized BVMSW exists twice in the onion state. Absorption in the two side arms contributes *independently* to the resonant feature that we measure at the eigenfrequency of mode A in the spectrum of Fig. 8(a). This will be important when we address localized spin waves in asymmetric rings discussed in Section 6. (In Section 5.4 we will discuss spin waves also for the case of a homogeneous $H_{\rm int}$. These spin waves extend around the ring in azimuthal direction.)

We convert the angular segment Φ_0 between the two turning points in Fig. 9(a) into a localization length using $\xi = \bar{R} \cdot \Phi_0$[rad]. We obtain $\xi^{\rm SA}_{(0,0)} \approx 1200$ nm for mode A(m, n)=A(0,0). In Figs. 9(c) and 9(d) we repeat the results from the micromagnetic calculation for modes A(0,0) and A(0,2) of the ring with w = 300 nm and compare them with the mode profiles based on the approximate analytical formalism. For the ring with

$w = 300$ nm the simulation yields $\xi^{\rm MS}_{(0,0)} \approx 1340$ nm. Here, the turning points are taken to be the points where the amplitude has dropped to 10% of the maximum amplitude, which is also the baseline value of the computed profiles. Both values, from the simulation and from the semi-analytical approach [Figs. 9(c) and 9(d), respectively] are in good agreement. Following the latter approach the localization length of mode $A(0,2)$ is $\xi^{\rm SA}_{(0,2)} \approx 1800$ nm. The simulation provides $\xi^{\rm MS}_{(0,2)} = 1780$ nm. These two values are again consistent. The larger localization length $\xi_{(0,2)}$ is a consequence of the lower frequency of this mode. The satellite modes seen in the experiment in Fig. 8(a) are thus higher order modes of mode A. Their eigenfrequencies decrease with increasing order since the excitations are BVMSWs obeying a negative group velocity for dipolar modes at small wave-vector k.

The calculation presented so far explains the mode localization of mode A in the side arm of a narrow ring. For mode B the formalism outlined in Section 4.3 would become involved. However, micromagnetic simulations presented in Fig. 6 have already shown that modes of type B in narrow rings are centered in the head and tail segments. They are localized DE-like spin waves. For them, the argumentation developed for modes of type A can be adapted as follows. In the head and tail segments at $\varphi = 0$ and π the internal field is at minimum. Away from these positions $H_{\rm int}$ increases. To stay at a fixed eigenfrequency a DE-like mode must exhibit a *smaller* wave-vector k_φ for *increasing* $H_{\rm int}$. Then $k_\varphi = 0$ defines a turning point beyond which the mode gets damped. By this means modes of type B are localized at minimum $H_{\rm int}$.

5.3. *From narrow to wide rings in the onion state: Transition from localized to extended modes*

In Fig. 6 we have depicted spin wave patterns belonging to modes A and B for different widths w. As the limiting case of a vanishing inner hole the microdisk is also shown. The patterns vary systematically with increasing width w. In Fig. 6 we have however not outlined how the eigenfrequencies of the spin wave excitations evolve with w. This will be discussed now. We have fabricated several arrays of rings and disks where we keep the outer diameter constant at 2.1 μm and vary the inner hole diameter $2R_{\rm i}$ from 1400 nm (narrow ring) down to 200 nm. Each array consists of nominally identical micromagnets and is integrated to a CPW. Spectra measured at 90 mT in the onion state of different rings are depicted in Fig. 10(a) to

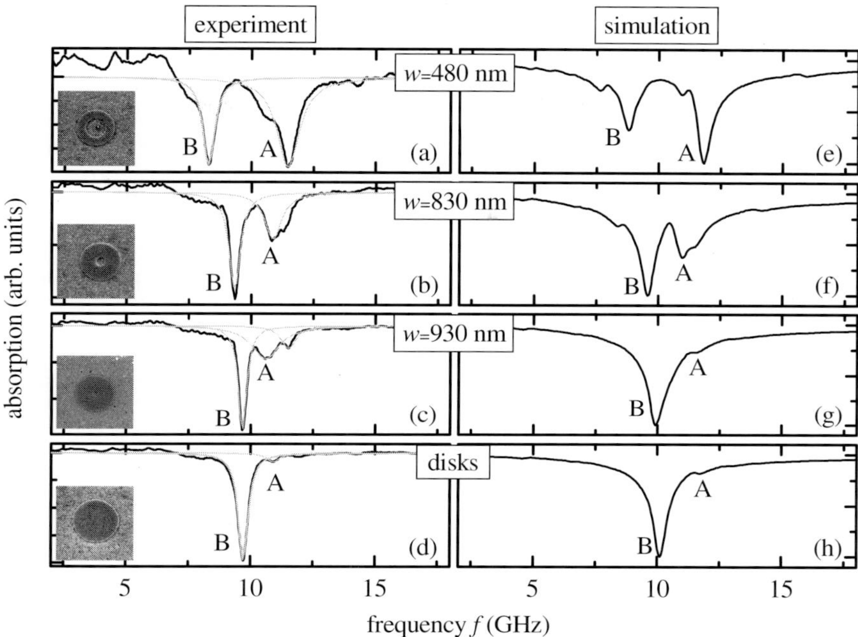

Fig. 10. Experimental spectra obtained on arrays with (a)–(c) permalloy rings with increasing w (see labels) and (d) disks. Graphs in (e) to (h) show the simulated spectra. Outer diameters are $2R_o = 2.1$ μm. The external field is $\mu_0 H = 90$ mT. Labels A and B indicate the characteristic main modes.

10(c). In Fig. 10(d) the spectrum is shown that we obtained on microdisks of the same R_o. We find that the eigenfrequencies and the relative absorption strengths evolve systematically as a function of w: starting from the narrow rings mode A is observed to shift to smaller frequencies. At the same time the absorption strength decreases. For mode B the opposite characteristics are observed. As the frequency difference between mode A and B shrinks the number of satellite modes decreases that we resolve. The eigenfrequencies and systematics observed in the experiments are in good agreement with the micromagnetic simulations depicted in Figs. 10(d) to 10(h).

In the following we explain the frequency shifts of the main modes A and B with w. For this we present a fully analytical approach that allows us to demonstrate that the dominant effect originates from the static and dynamic demagnetization fields. Satellite modes, i.e., higher order modes where the finite spin-wave wave-vector k becomes larger and gains further importance, are not described by the following ansatz. This is because the analytical modeling assumes uniform spin precession. The approach can clarify however the relative dependence of the main modes on width w. For a thin film, Kittel calculated the resonance frequency a long time ago.[48] Since then the formalism has entered solid state physics textbooks. For a large plain film it is assumed that all spins precess at the same frequency and that the wave-vector k is zero. This uniform excitation is called FMR as outlined in Section 2. Boundaries of the film introduced demagnetization fields H_{dm}. These are large if spins point perpendicular to the edge and small if spins are parallel. In particular it is shown that both types of demagnetization fields, i.e., static and dynamical fields, are relevant to explain FMR eigenfrequencies. In the following we adopt the Kittel formalism[48] for a thin and infinitely long wire of width w. Here we obtain an FMR eigenfrequency according to

$$f_{\mathrm{res}} = \frac{\gamma\mu_0}{2\pi}\Big((H + (N_y - N_x)M_S)\cdot(H + (N_z - N_x)M_S)\Big)^{\frac{1}{2}}. \qquad (16)$$

For the in-plane direction we use the demagnetization factors N_x and N_y. The x and y axes are defined in Figs. 11(a) and 11(b). $\vec{H}$ is applied in the x direction in Fig. 11. N_z accounts for the direction perpendicular to the plane. For a wire oriented along, both, the direction of the external field H and the x axis [Fig. 11(a)] the demagnetization factors in the center of the magnet are: $N_x = 0$, $N_y = \frac{2t}{\pi w}$ and $N_z = 1 - \frac{2t}{\pi w}$.[46] Such values of N_y and N_z lead to positive prefactors of M_S in Eq. (16). By this means the precession frequency f_{res} is increased if compared to an unpatterned film. N_x and N_y for a wire oriented along the y axis [Fig. 11(b)], i.e., perpendicular to the magnetic field direction, are calculated to be $N_x = \frac{2t}{\pi w}$, $N_y = 0$ and $N_z = 1 - \frac{2t}{\pi w}$. A negative prefactor results for M_S in Eq. (16) and, consequently, H_{int} and f_{res} are reduced. In Fig. 11(c) we compare the two different eigenfrequencies of long permalloy wires as a function of w (full lines). Strikingly, the two main modes A and B of rings (closed symbols) follow the two FMR modes calculated for wires of the same width which are magnetized in the two orthogonal directions, i.e., along and perpendicular to the wire axis. The simulated eigenfrequencies (open symbols) follow both the experimental data and the calculated curves. Only for mode A and $w \geq 900$ nm a discrepancy occurs. The characteristic dependencies on w demonstrate that demagnetization fields due to the mesoscopic width are important for the spin wave eigenfrequencies. Rings in the onion state exhibit in fact two extreme spin orientations: spins point perpendicular to

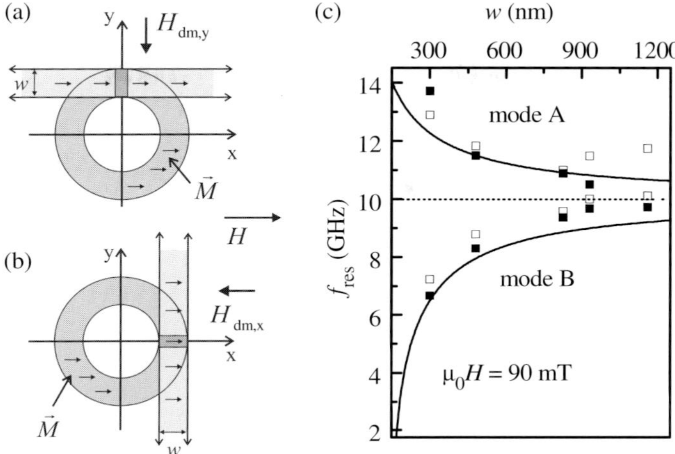

Fig. 11. (a) Sketch of a ring with a narrow wire which is aligned with the direction of H (reproduced from Fig. 2). If spins are excited and deviate from the x direction a *dynamical* demagnetization field $H_{\text{dm},y}$ results. This increases the eigenfrequency if compared to an unpatterned film.(b) Ring and a narrow wire oriented perpendicular to H. The magnetization points towards to the wire edges and thereby generates a large *static* demagnetization field $H_{\text{dm},x}$. This reduces the internal field H_{int} below H and lowers the spin precession eigenfrequency if compared to (a). (c) Eigenfrequencies of the main modes A and B vs width w: closed symbols — measured data, open symbols — data extracted from micromagnetic simulations like the ones shown in Figs. 6 and 10, full lines — eigenfrequencies according to Eq. (16) calculated for the two scenarios shown in (a) [lower curve] and (b) [upper curve], dotted line — eigenfrequency of the uniform precession mode in the unpatterned permalloy film.

the edges in the head and tail and parallel to the ring's boundary in the ring arms [cf. Figs. 5 and 11(a) and 11(b), respectively]. The qualitative comparison with wires treated by the analytical Kittel formalism highlights that spins in the different ring segments precess at different eigenfrequencies. This is due to the inhomogeneous internal field that varies in azimuthal direction. A full quantitative comparison should not be made, however, since Eq. (16) neglects the finite wave-vector $\vec{k}$ due to effective pinning at the boundary[36] and localization along the azimuthal direction. Realistic mode patterns have been imaged on Co rings in Ref. 58 and, for permalloy rings, are illustrated in Fig. 6. These graphs demonstrate that the spin excitations evolve from localized (on the left) to extended modes (on the right) as the width w increases. This transition[27] is not considered by the Kittel formalism of wires.

5.4. *Rings in the vortex state: Spin wave interference and azimuthal modes*

We now turn to spin excitations in the vortex state. At $H = 0$ all spins are aligned with each other and point along the edges of the ring such that H_{int} is constant. Interestingly, applying a finite in-plane field with $|H| < |H_1^{\text{sw}}|$ generates a polarized vortex state exhibiting an internal field $H_{\text{int}}(\varphi)$ that oscillates smoothly around the ring. In the center column of Fig. 5 we compare micromagnetic simulations of the internal field of the onion (top) and vortex (bottom) state at non-zero in-plane field. We found that $H_{\text{int}}(\varphi) \approx H \cdot \sin\varphi$ models well H_{int} of the vortex state. In this internal field landscape we have recently observed two types of spin wave modes: (i) mode A′ which resembles mode A already discussed for the onion state of rings and (ii), in particular, modes I_n.[63] We find that for narrow rings the magnetic field dispersions of modes A and A′ are identical. Only for $w > 400$ nm there is a frequency offset between them (cf. Fig. 12). Mode A′ resides in the ring arm, where $\vec{M}$ is aligned with $\vec{H}$, and is localized in the same segment as mode A (cf. upper and lower graph in the right column of Fig. 5). In the following we will show that spin waves I_n circulate around the ring with a characteristic wave-vector $k_\varphi = k_{n\varphi}$ and form azimuthal modes with a number n of nodal lines. Intriguingly, spin wave excitations I_n with an eigenfrequency smaller than mode A′ can span the whole ring. A ferromagnetic ring in the vortex state is found to act as a ring resonator for BVSMWs. In Fig. 12 we summarize, as a gray-scale plot, absorption spectra taken on rings with $w = 600$ nm at different in-plane fields $\mu_0 H$ ranging from -60 mT to $+60$ mT. Black color encodes a strong power absorption and highlights the measured spin wave eigenfrequencies. At each field the spectrum was recorded *after* applying a saturation field of $+90$ mT. Thus we follow the magnetic field dependence of spin wave modes. The rings exhibit the typical reversal behavior with two switching fields. We find $H_1^{\text{sw}} = -2$ mT and $H_2^{\text{sw}} = -18$ mT in this case. In the vortex state ($H_2^{\text{sw}} < H < H_1^{\text{sw}}$) the intriguing novel observation is the step-wise behavior of the low-frequency branch (labeled by I_n with $n = 1, 2, 3, 4$). Here, spin-wave eigenfrequencies are discrete and form plateaus I_n. Clear frequency gaps of $\cong 0.5$ GHz are resolved between the plateaus I_n which are found to be slightly magnetic field dependent. This interesting behavior is best seen in Fig. 13(b) where we present a detailed measurement performed in a minor loop. For this, we have first saturated the rings at $+90$ mT and second applied $\mu_0 H = -14$ mT (cf. H_{ML} in Fig. 12) thus generating the

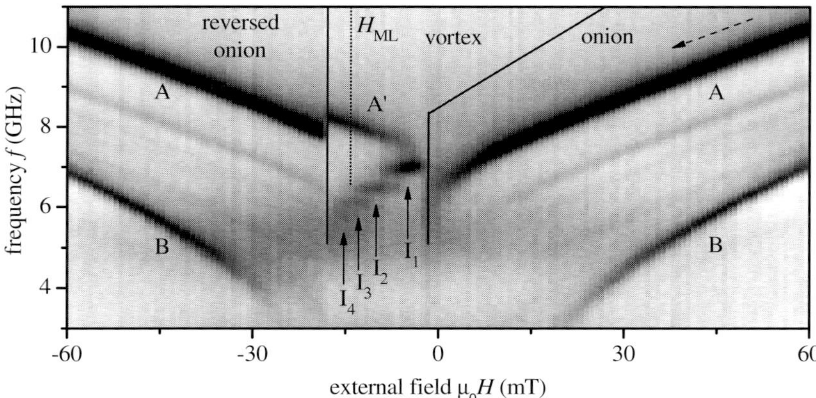

Fig. 12. Gray-scale plot of absorption spectra taken at successively decreased magnetic field (cf. dashed arrow) after saturation at $\mu_0 H = +90$ mT (onion state). The ring parameters are: $w = 600$ nm, $t = 30$ nm, $2R_o = 1950$ nm. Dark represents strong absorption. In the onion state modes A and B are detected consistent with Section 5.1. In the vortex state discrete spin wave eigenfrequencies I_1, I_2, I_3, I_4 (see labels) and A′ are resolved. Vertical full lines indicate switching fields. The vertical dotted line depicts $\mu_0 H_{ML} = -14$ mT. The minor loop in Fig. 13(b) is for $H > H_{ML}$.

vortex state. The regime of the vortex state is identified easily: the spin wave modes show a mirror symmetry with respect to $H = 0$ consistent with Ref. 29. In Fig. 13(a) spin excitations in the vortex state of a further ring sample are depicted. Again, we observe gaps in the low-frequency branch. They are less pronounced since the geometrical parameters are different. This will be explained in detail later.

In Figs. 13(a) and 13(b) only mode I_1 is resolved at $H = 0$. In 13(b) the eigenfrequency is $f = 7.04$ GHz. Experimentally we find that each of the spin-wave excitations I_n is detected within a characteristic field interval. The onset fields of modes I_n with $n = 2, 3$ and 4 are 4, 8 and 9 mT, respectively. The modes exist up to a critical field $H_{c,n}$: we find $\mu_0 H_{c,1} =$ 9 mT for mode I_1, $\mu_0 H_{c,2} = 13$ mT and $\mu_0 H_{c,3} \approx 15$ mT. The critical field (eigenfrequency) is found to increase (decrease) with increasing index n. The mode A′ in Fig. 13(a) [13(b)] does not exist for $\mu_0|H| < 2.5$ mT [$\mu_0|H| < 4$ mT] and, in contrast to modes I_n, exhibits in particular no steps. The magnetic field dispersion is opposite to the one of modes I.

For the discussion let us revisit the semi-analytical approach of Section 4.3. To calculate the spin dynamics of rings in the vortex state we now argue that the ring's azimuthal vector k_φ is quantized by the

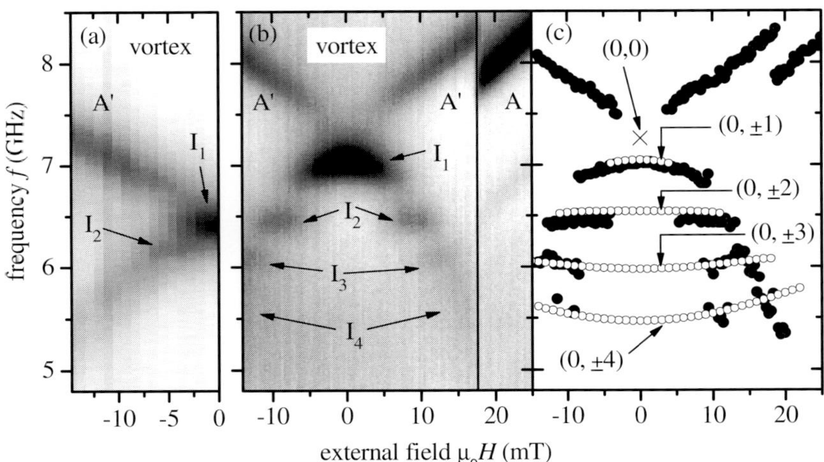

Fig. 13. (a) Spin wave excitations in the vortex state of rings with $w = 500$, $t = 25$ nm and $2R_o = 2150$ nm. Dark represents strong absorption. (b) Spectra taken in the minor loop starting at $\mu_0 H_{\rm ML} = -14$ mT (cf. Fig. 12) for the rings with $w = 600$ nm, $t = 30$ nm and $2R_o = 1950$ nm. In the vortex state modes have a mirror symmetry with respect to $H = 0$.[29] The vertical line indicates switching to the onion state at $\mu_0 H \geq 18$ mT. (c) Full symbols are eigenfrequencies extracted from (b). Open symbols refer to the quantized modes calculated via Eq. (17). Numbers in brackets label the modes (m, n) from the modeling.

condition:

$$n \cdot 2\pi = \oint k_\varphi[f, H_{\rm int}(\varphi)]\ \rho\ {\rm d}\varphi. \tag{17}$$

ρ is the radius of the integration path. Equation (17) is the condition for constructive interference of spin waves propagating around a ring. For $H \neq 0$, we have shown in Fig. 5 that $H_{\rm int}(\varphi)$ oscillates around the ring. The wave vector $k_\varphi(\varphi) = k_\varphi[f, H_{\rm int}(\varphi)]$ thus varies as a function of φ. In the following we recalculate the eigenfrequencies using the quantization condition Eq. (17). Incorporating Eq. (17) in the semi-analytical approach allows us in particular to obtain the field dependence of eigenfrequencies f_n.

If we neglect the exchange energy in the remanent vortex state the ring is mathematically equivalent to a longitudinally magnetized wire which exhibits *no* demagnetization field *at the front and back end*. We can use the dispersion relations $f(\kappa, H_{\rm int})$ of Ref. 34 to calculate spin-wave excitations for a longitudinally magnetized wire forming a ring if we consider periodic boundary conditions at the two wire ends. The most important difference

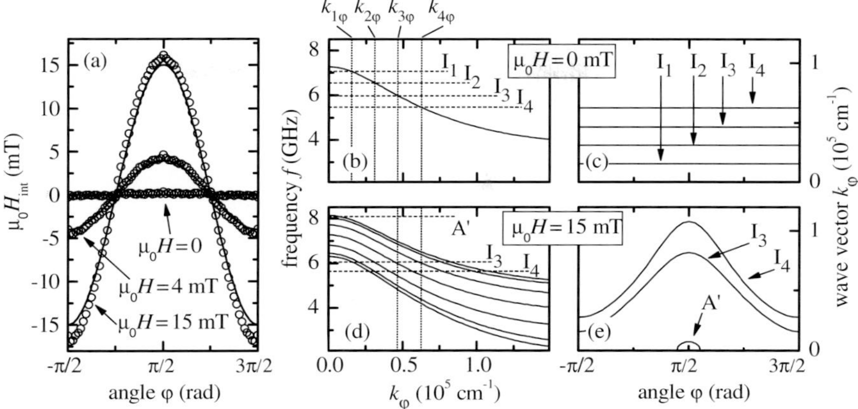

Fig. 14. (a) Internal field simulated for the 600 nm wide permalloy rings at three different H (symbols): H_{int} is depicted at the position $\rho = \bar{R}$ as a function of φ. The full lines correspond to $H_{\text{int}} = H \cdot \sin\varphi$. (b) and (d) show wave vector dispersions valid at distance $\bar{R}$ from the ring's center at different angles φ for $H = 0$ and $\mu_0 H = 15$ mT, respectively. Dispersions in (d) are calculated at seven distinct values φ. (c) and (e) depict angular dependencies of wave vectors for modes I_n and A′.

to the straight wire occurs if we apply a magnetic field H to the ring: H_{int} becomes a spatially oscillating function (cf. Fig. 5). For the following calculations we focus on the azimuthal direction $\vec{e}_\varphi$ at $\rho = \bar{R} = R_{\text{o}} - w/2$, i.e. the central radius. Open symbols in Fig. 14(a) refer to a micromagnetic simulation of the internal field for the 600 nm wide rings that we investigated experimentally in Fig. 12. The functional form $H_{\text{int}} = H \cdot \sin\varphi$ (solid lines) models the simulated internal field very well. It provides a good approximation at small field H where deviations from the circular symmetry of M (cf. left bottom column in Fig. 5) are negligible. Considering the characteristic H_{int}, we calculate the wave-vector dispersion $f[k_{m\rho}, k_\varphi, H_{\text{int}}(\varphi)]$ using Eqs. (7), (10) and (11) at different φ from the semi-analytical approach of Ref. 34. Results are shown in Figs. 14(b) and 14(d) for $H = 0$ and $\mu_0 H = 15$ mT, respectively. For the calculations the exact parameters are $2R = 1920$ nm, $w = 630$ nm and $t = 26$ nm, consistent with the AFM measurements within the experimental accuracy. The saturation magnetization $\mu_0 M_S = 1.12$ T is provided by FMR measurements on an unpatterned permalloy film. Introducing wave-vectors $k_\varphi[H_{\text{int}}(\varphi)]$ at $\rho = \bar{R}$ into Eq. (17), i.e., the condition for constructive interference, then defines the discrete eigenfrequencies of the vortex state.

To discuss the quantization phenomenon in detail it is instructive to start with $H = 0$ (remanent vortex state) where the internal field is zero everywhere and the dispersion of the BVMSW is not position dependent. This interesting case is shown in Fig. 14(b) where we assume for the radial direction $m = 0$, i.e., $k_{m\rho} = \pi/w_{\text{eff}}$. This corresponds to zero nodal lines in the radial direction. The wave-vectors $k_{n\varphi}$ consistent with Eq. (17) are constant around the ring [Fig. 14(c)] and are given by $k_{n\varphi} = n/\bar{R}$. The corresponding frequencies labeled by I_n exhibit the experimentally observed gaps of 0.5 GHz. The allowed modes in the vortex are indexed by $(m, \pm n)$ where the sign accounts for the direction of BVMSW propagation, i.e., clockwise or counterclockwise around the ring. The excited BVMSWs are guided around the ring and interfere.

For the ring at a finite field we obtain a band of dispersions as the wave-vectors of the BVMSW becomes position dependent. The case for $\mu_0 H = 15$ mT is shown in Fig. 14(d). The upper and lower 'band edges' are given by the maximum and minimum H_{int} at $\varphi = \pi/2$ and $\varphi = 3\pi/2$, respectively. The band broadens around the dispersion of $H_{\text{int}} = 0$. Graphically, wave-vectors $k_{n\varphi}(\varphi)$ are found as intersections of the horizontal frequency lines I_n with the dispersions. The wave vectors which are consistent with Eq. (17) are depicted in Fig. 14(e). In particular they now vary with φ. Due to the broadening of the dispersion band with applied field the calculated eigenfrequencies f_n depend on H. This is best seen in Fig. 13(c) where we compare the calculated f_n (open symbols) with the measured eigenfrequencies (filled symbols) as a function of H. Though small, the field dependence is clearly visible. It originates from the broadening of dispersion curves that is not perfectly symmetric around the spin wave dispersion at H_{int}. Intriguingly spin wave excitations with eigenfrequencies I_n can intersect all of the dispersion curves [cf. also Fig. 5] and span over the whole ring. This is in contrast to the mode A′. The dispersion band at $\mu_0 H = 15$ mT in Fig. 14(d) also explains why the spin wave excitations I_n are observed only below an upper critical field. To exist everywhere in the ring the horizontal frequency lines I_n must, first, intersect *all* dispersion relations of the band and, second, accomplish Eq. (17). These criteria are fulfilled for I_3 and I_4 at $\mu_0 H = 15$ mT. Below $k_{3\varphi}(\varphi)$ in Fig. 14(d) no further wave-vectors are found which are consistent with both requirements. BVMSW modes extending around the ring with $n = 1$ and $n = 2$ do not exist. The model correctly predicts upper critical fields $H_{c,n}$ that increase with increasing mode number n. The calculated modes shown as open symbols in Fig. 13(c) exist over the whole interval $[-H_{c,n}, +H_{c,n}]$, whereas we observe onset fields for $n \geq 2$ in the experiment. The semi-analytical approach however does not include the specific excitation mechanism. The high-frequency field $h_{\rm rf}$ is homogeneous in the rings' plane. To excite BVMSW of high wave-vectors $k_{n\varphi}$ with $n \geq 2$ an inhomogeneous internal field $H_{\rm int}$ or a spatially inhomogeneous magnetization are needed. To generate this, we have to apply a finite field leading to the experimentally observed onset field. Our modeling predicts an eigenfrequency $f = 7.26$ GHz of the uniform azimuthal mode with $k_\varphi = 0$ at $H = 0$. This mode $(m, n) = (0, 0)$ is marked by the cross in Fig. 13(c). Following Refs. 77, 78 the excitation of modes with $n = 0$ is forbidden due to both the symmetry of the vortex state and the orientation of $h_{\rm rf}$.

To explain mode A′ in the vortex state we start from the measured frequency $f = 8.1$ GHz at $\mu_0 H = 15$ mT. In Fig. 14(d) this frequency lies at the upper edge of the dispersion band. This corresponds to the ring segment at $\varphi = \pi/2$ where the magnetization is parallel to H and $H_{\rm int}$ is large. This has also been illustrated in Fig. 5 (bottom row) and compared to mode A in the onion state. $k_\varphi(\varphi)$ shown in Fig. 14(e) suggests that mode A′ is localized near $\varphi = \pi/2$ since real values for the wave-vector exist only for $\pi/2 - \frac{\pi}{10} < \varphi < \pi/2 + \frac{\pi}{10}$. The corresponding localization length amounts to 420 nm. Due to the localization spin-wave interference around the ring is not possible for mode A′ and discrete steps are not expected in the magnetic field dispersion. This is in agreement with the experimental data. Mode A′ is similar to mode A in the onion state which was discussed in Refs. 28, 29, 63. As the magnetization pattern and the internal field profile are not identical for the vortex and the onion state one might expect that this characteristic spin wave excitation shifts to a different eigenfrequency. We observe a frequency shift on our permalloy rings [cf. Fig. 13(b)], however, only for $w > 400$ nm.

In Fig. 13(a) we find that for rings with $w = 500$ nm and $t = 25$ nm the frequency gap between I_1 and I_2 is smaller than for the rings with $w = 600$ nm and $t = 30$ nm shown in Fig. 13(b). Considering the different geometrical parameters in the semi-analytical approach confirms the reduced gaps. In particular, the negative slope of the spin wave dispersion $f(k_\varphi)$ is calculated to be less steep if $w = 500$ nm. The change in slope results in smaller gaps for modes I_n fulfilling Eq. (17). For the rings which we will discuss in Section 6 the frequency gaps in the lower branch of the vortex state will not be resolved at all and *both* branches A′ and I_n will appear to be continuous.

The eigenfrequencies f_n shown as open symbols in Fig. 13(c) are provided by Eq. (17) which we evaluated exactly at $\rho = \bar{R}$. This is a one-dimensional ansatz and might be considered a drawback of the semi-analytical approach. In general, the spin wave amplitude of a mode with $m = 0$ is non-zero for $R_\mathrm{i} < \rho < R_\mathrm{o}$ in radial direction. However dipolar pinning occurs at the edges of a permalloy nanomagnet[36] and the amplitude vanishes when approaching the inner and outer edge of the ring. By the dipolar pinning the spin wave with $m = 0$ exhibits a maximum amplitude in the region near $\rho = \bar{R}$. This might explain the good agreement between experiment and semi-analytical approach considering only $\rho = \bar{R}$.

6. Spin Wave Spectroscopy on Asymmetric Rings

In this section we study rings with a decentered inner hole. We call such devices in the following asymmetric rings. We focus here on the effect of the hole displacement a on the spin wave excitations. We present and discuss data obtained on two arrays of different permalloy rings. Again we integrate the arrays on coplanar waveguides. For sample 1 (Section 6.1) the asymmetry is provided by an inner hole of diameter $d_i = 2R_\mathrm{o} - w1 - w2 =$ 1040 nm that is offset by $a = 65$ nm as measured by the AFM [cf. Fig. 15(c) for the definition of parameters]. The outer diameter $2R_\mathrm{o}$ is 2.2 μm. The thickness t amounts to 25 ± 5 nm. The hole is offset perpendicular to the CPW direction. This defines the mirror symmetry axis as depicted in Fig. 15(c). In the arrays the center-to-center spacing is 4 μm to minimize magnetic coupling. For sample 2 (Section 6.2) the parameters are: $d_i =$ 1100 nm, $2R_\mathrm{o} = 2.01$ μm, $t = 35 \pm 5$ nm, $a = 180$ nm. The asymmetry of sample 2 is intentionally larger if compared to sample 1. For all values in lateral direction the AFM accuracy is ± 20 nm.

6.1. *Rings with a small asymmetry: Frequency splitting of localized modes*

Let us first discuss spectra measured on the ring array with a small asymmetry (sample 1). The small asymmetry does not lead to a severe change in the quasistatic properties, i.e., one still finds the onion and the vortex state. However switching characteristics are expected to change. If one applies a field $\vec{H}$ perpendicular to the mirror axis one anticipates two different switching fields H_2^{sw} for the vortex-to-reversed-onion transition depending

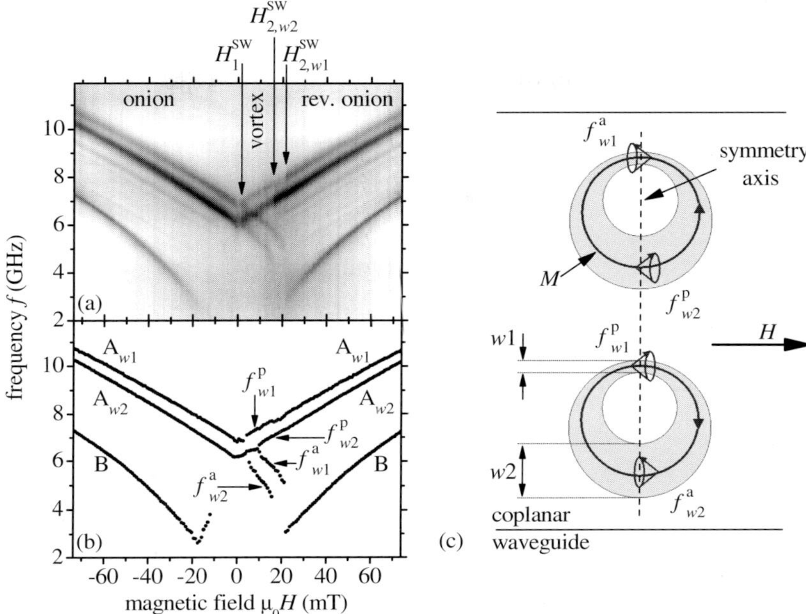

Fig. 15. (a) Gray-scale plot summarizing absorption spectra measured an rings with a small asymmetry ($a = 65$ nm). The field H is applied along the CPW and perpendicular to the rings' mirror axis. (b) Eigenfrequencies extracted from the main modes of (a). The labeling refers to the different widths $w1$ and $w2$, and different orientations of the local $\vec{M}$ with respect to $\vec{H}$ (parallel vs antiparallel). (c) Sketch of asymmetric rings integrated on a coplanar waveguide. The offset a of the inner hole with respect to the ring's center coordinate is given by $a = (w2 - w1)/2$. The clockwise (bottom) and counterclockwise (top) vortex circulation directions are illustrated. From these, four different combinations of ring arm widths and relative ($\vec{M}$,$\vec{H}$) orientations result for the segments highlighted by the spin precession cones.

on the circulation direction of the vortex state. They result from the different widths $w1$ and $w2$ [cf. Fig. 15(c)].[66] If in an array of asymmetric rings the vortex circulation direction is non-deterministic and both directions are present one therefore expects to observe the two different switching fields $H^{\rm sw}_{2,w1}$ and $H^{\rm sw}_{2,w2}$.

In Fig. 15(a) we show a gray-color plot of absorption spectra obtained on sample 1. We do observe irreversible switching at three distinct fields $H^{\rm sw}_1$, $H^{\rm sw}_{2,w1}$ and $H^{\rm sw}_{2,w2}$ in agreement with the expectation. In Fig. 15(b) we extract the eigenfrequencies of the main modes. The important observation is that three main modes A_{w1}, A_{w2} and B appear in the onion state, instead of the two modes A and B in symmetric rings. In the latter case, spin waves of type A are confined to two separate ring halves (cf. Fig. 6). The eigenfrequencies f_{A} of the two ring halves are degenerate due to the same width w. We find that such spin excitations of type A split in two separate eigenfrequencies A_{w1} and A_{w2} in the asymmetric rings. This results from the ring's width w which, intentionally, is no longer constant as a function of azimuthal angle φ as shown in Fig. 15(c). According to the results in Section 5.3 the different widths $w1$ and $w2$ generate different eigenfrequencies due to changes in the dynamical demagnetization fields and radial wave-vector k_ρ. In the onion state at $H < 0$, the modes A_{w1} and A_{w2} are attributed to the narrow and wide side arm, respectively. The absorption strengths differ significantly. In Fig. 15(a) the signal of A_{w2} is large compared to A_{w1} because more spins are excited in the wide side arm. We have found the same characteristics for sample 2 (Fig. 16). Our findings are confirmed by micromagnetic simulations. The spectrum calculated for the onion state of an asymmetric ring of sample 2 is shown in Fig. 16(c). The agreement with the experimental spectrum depicted in Fig. 16(b) is rather good. Spin wave pattern in Figs. 16(i) and (ii) show convincingly that the modes at eigenfrequencies A_{w1} and A_{w2} are confined in opposing ring segments where $\vec{M}$ is parallel to both the field $\vec{H}$ and the ring's edges [Fig. 16(iv)]. Also mode B which resides in the domain walls is modeled well. The spin wave pattern in Fig. 16(iii) highlights that here the two localized spin-wave excitations are still degenerate and exhibit the same eigenfrequency. In the following we will focus on the peculiar spin dynamics properties of asymmetric rings and will no longer discuss mode B.

When the asymmetric rings enter the vortex state at H_1^{sw} we observe the occurrence of four modes in Fig. 15(b). Two of them exhibit a positive magnetic field dispersion and the other two a negative one. For the discussion let us assume that the array consists of asymmetric rings with a statistical distribution of clockwise and counterclockwise vortex circulation directions. The two configurations are depicted in Fig. 15(c). In this case the four spin-wave eigenfrequencies are explained by four different combinations of ring arm widths and relative ($\vec{M}$,$\vec{H}$) orientations [Fig. 15(c)]: f^p_{w1} and f^a_{w1} (narrow ring arms, $\vec{M}$ parallel and antiparallel to $\vec{H}$), f^p_{w2} and f^a_{w2} (wide ring arms, $\vec{M}$ parallel and antiparallel to $\vec{H}$, respectively). Each of these modes exhibits a characteristic magnetic field dispersion reflecting the local orientation of $\vec{M}$ with respect to $\vec{H}$. The important point here is that a positive (negative) sign of the magnetic field dispersion tells us whether the local magnetization $\vec{M}$ of a ring segment is parallel

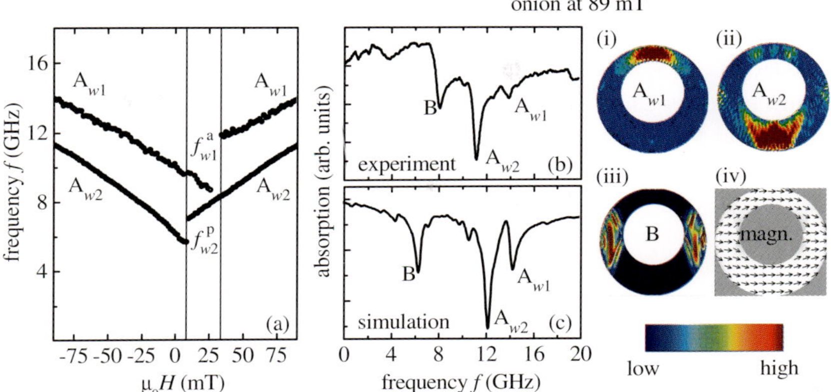

Fig. 16. (a) Eigenfrequencies extracted from absorption spectra taken on the array containing rings with a large asymmetry $a = 180$ nm (sample 2). Mode B is not depicted. In the vortex state, that is between the two vertical lines, only *two* instead of *four* spin wave resonances are found. This means that the vortex circulation direction of the rings is the same. The field H is applied along the CPW, i.e., perpendicular to the mirror axis. (b) Experimental and (c) simulated spectrum in the onion state at $\mu_0 H = 89$ mT. The graphs (i) to (iii) on the right hand side show spin wave patterns, graph (iv) depicts the static magnetization configuration at 89 mT. Small arrows highlight the orientation of the local $\vec{M}$.

(antiparallel) with respect to $\vec{H}$. For a single ring in the vortex state the absorption strength at eigenfrequency f^p_{w2} should be larger than f^p_{w1} and should scale like $w2/w1$. This is consistent with the data of sample 1. For an array the relative mode intensities are in addition determined by the exact number of ring arms that contribute to one of the four possible eigenfrequencies. In the extreme case that an array exhibits a coherent vortex circulation direction, i.e., it is magnetically ordered, two modes should be *absent* in the vortex state since the corresponding ($\vec{M}$,$\vec{H}$) orientations would be missing. This will be the case for sample 2 where we will focus exclusively on the vortex state in the following.

6.2. *Rings with large asymmetry: Deterministic control of the vortex circulation*

In the experimental data of Fig. 16(a) we observe that in the vortex state (i) there exist only two spin-wave eigenfrequencies. Within the experimental resolution, the modes f^a_{w2} and f^p_{w1} found in Fig. 15(b) are missing. (ii) The absorption strength (not shown) of mode f^p_{w2} is found to be constant at the switching field of $\mu_0 H_2^{sw} = +29$ mT where the rings enter the reverse onion state. A constant absorption strength is also true for f_{w1}^a at $\mu_0 H_1^{sw} = +8$ mT where the rings enter the vortex state. Spin-wave eigenfrequencies characteristic for rings with clockwise vortex circulation direction are missing in the experimental data. The eigenfrequencies and field dispersions observed in Fig. 16(a) suggest an *ordered* vortex state in the array consisting of *counterclockwise* vortex circulation directions only. Note that in this experiment, we use a magnetic field history with $\vec{H}_{\rm his}$ applied *perpendicular* to the symmetry axis, i.e., along the CPW. A detailed minor-loop measurement (not shown) on the same array with $\vec{H}_{\rm his}$ being *parallel* to the symmetry axis reveals for the vortex state the complete set of four modes f_{w1}^p, f_{w1}^a, f_{w2}^p and f_{w2}^a.[26] We get equal absorption strengths suggesting that the fractions of clockwise and counterclockwise circulation directions are roughly equal. This shows that depending on $\vec{H}_{\rm his}$ we are able to generate both an ordered and disordered array configuration in the vortex state.

Let us address the mechanism for vortex circulation control within the array only briefly. We have argued that in symmetric rings the transition from onion to vortex state is mediated by domain wall propagation. Micromagnetic simulations for our rings with large asymmetry reveal that the reversal process is now different: for $\vec{H}$ parallel to the CPW the wide ring segment of width $w2$ reverses via buckling. The subsequent switching by nucleation leads to a small switching field $H_{2,w2}^{\rm sw}$ and dominates the transition to the vortex state with deterministic circulation direction.[26] This deterministic control of the vortex circulation is essential if rings are to be used as building block for future magnetoelectronics or magnonics applications.

7. Nonlinear Phenomena in Rings

We have investigated nonlinear spin wave phenomena in symmetric rings exhibiting a width $w = 700$ nm. At a low power level of the VNA of -12 dBm (0.063 mW) in Figs. 17(a) to 17(c) (uppermost curves) the rings display spin wave modes consistent with our discussion in Sections 5.3 and 5.4. For the following experiments we have increased the power level in a stepwise manner to $P = +8$ dBm (6.3 mW). To observe nonlinear phenomena the amplitude of $\vec{h}_{\rm rf}$ at the rings' position on the CPW should be as large as possible. To generate a high amplitude $\mu_0 h_{\rm rf}$ of 3.5 mT at $P = +8$ dBm

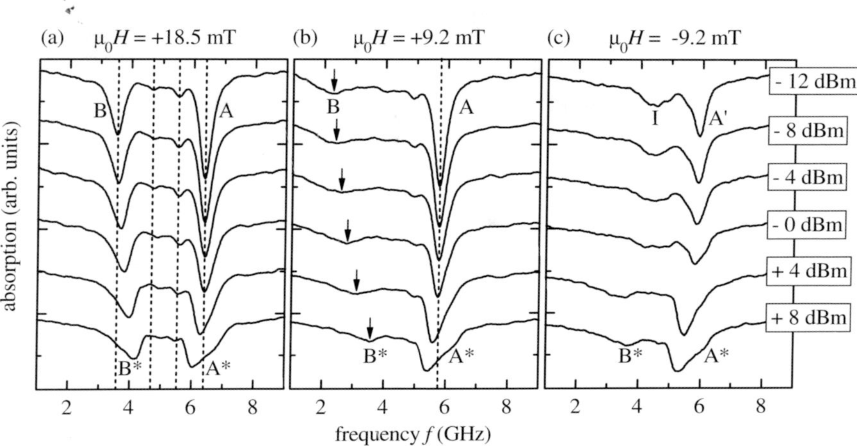

Fig. 17. Experimental spectra taken in the onion state of symmetric rings with $2R_o = 2.25$ μm, $w = 700$ nm and $t = 25$ nm at (a) $\mu_0 H = +18.5$ mT and (b) $\mu_0 H = +9.2$ mT. Individual spectra are offset for clarity. From top to bottom the output power P of the VNA is increased as indicated on the right. The modes are found to shift characteristically with P. The rings remain in the onion state. At high power the main modes of the onion state are labeled by A* and B^*. (c) Experimental spectra taken at $\mu_0 H = -9.2$ mT. The spectrum on the top reflects the vortex state of the rings, i.e., we observe the main modes A′ and I already discussed in Section 12. Also these modes shift with increasing power. At the critical $P_c = +4$ dBm the rings are however no longer in the vortex state but have switched to the onion state [compare to (b)]. Due to this microwave assisted switching we observe the modes A* and B* in the bottom trace.

we tapered the CPW's central conductor to a narrow width $w_C = 2.86$ μm. The small CPW accommodates 20 nominally identical rings. The rings are found to exhibit the typical reversal behavior with two switching fields: for $\mu_0 H \geq \mu_0 H_1^{\rm sw} = +5.6$ mT and $\mu_0 H \leq \mu_0 H_2^{\rm sw} = -10.6$ mT the rings are in the onion and reverse onion state, respectively, if the magnetic history is $\mu_0 H_{\rm his} = +90$ mT. The external field is applied along the CPW.

7.1. *Power dependence of spin wave modes in symmetric rings*

In Fig. 17(a) we show spin wave spectra obtained in the onion state at $\mu_0 H = +18.5$ mT. With increasing VNA output power P the spectra exhibit four striking changes:[64] (i) mode A (mode B) shifts to a *smaller* (larger) eigenfrequency f, (ii) the absorption strengths decrease and (iii) the line widths increase. (iv) For both types of modes the absorption features become asymmetric. However, the asymmetry is different for the two modes. We label these modes which are characteristic for the onion state in case of a large VNA output high power with A* and B*. We attribute the observed changes to nonlinear spin dynamics that are due to large spin-precession angles. For ferromagnetic resonances in a macroscopic magnet it was shown that the magnet's shape and orientation in an external field $\vec{H}$ that saturated the sample played an important role for the observed frequency shifts.[11,18,72] For example, the uniform mode of a macroscopic disk was found to shift to *smaller* (larger) f with increasing P if $\vec{H}$, i.e. $\vec{M}$, was oriented *parallel* (perpendicular) to the *easy plane*.[11] For our microscopic magnets we observe in Fig. 17(a) that the frequency of A becomes *smaller*. The observed characteristics in the nonlinear regime agree well with our explanation in Section 5.3 that mode A is localized in a ring segment where the magnetization M is parallel to the *easy direction*, i.e., the edges of the ring (cf. Fig. 6). Consistently we find a shift to larger f with P for mode B which resides in a segment where $\vec{M}$ is perpendicular to the edges. The exact asymmetry of the resonances in rings still needs to be elucidated. In Fig. 17(b) we show spin wave spectra obtained in the onion state at the smaller field $\mu_0 H = +9.2$ mT. We observe the same systematic variation of modes A and B. At high power P the modes become A* and B*.

7.2. *Microwave assisted switching*

We now turn to Fig. 17(c) where we first discuss the uppermost spectrum obtained at $P = -12$ dBm. The absorption is for $\mu_0 H = -9.2$ mT$<$ $H_1^{\rm sw} = -0.9$ mT in the vortex state. Consistent with our earlier discussion in Section 5.4 on spin wave interference phenomena we observe both the modes I_n that reflect azimuthal spin wave modes and the localized mode A′. When we now increase the power P (but stay below +4 dBm) we find five characteristic changes with increasing P: (i) modes A′ and I both first shift to *smaller* frequencies f, (ii) their intensities decrease and (iii) line widths increase. (iv) The resonances become asymmetric and (v) mode I is found to split up into further absorption lines I_n, each with a distinct power dependence. We attribute this regime below +4 dBm to the nonlinear spin dynamics as discussed in Section 7.1. At $P \geq +4$ dBm the absorption spectrum changes completely: we find modes A* and B* which are characteristic for the onion state. This is evidenced if the spectra at the bottom of Figs. 17(b) and 17(c) are directly compared. They are recorded intentionally at the same absolute field H but with different initial magnetization states. The spectra at $P = +8$ dBm are however the same. The sequence of spectra depicted in Fig. 17(c) thus confirms microwave assisted switching (MAS) from the (initial) vortex state to the reversed onion state. MAS is an intriguing phenomenon which currently attracts great interest in spin dynamics research.[64,73] Further experimental and theoretical work is necessary to understand details of the microscopic mechanism in mesoscopic magnets.

8. Conclusions

We have reported broadband GHz spectroscopy on nanostructured permalloy rings. This device geometry has turned out to be interesting for fundamental research on spin excitations in nanomagnets. Depending on the exact ring width and thickness, the external field and the initial quasi-static magnetization we have observed a variety of intriguing spin wave phenomena: localization in the ring's inhomogeneous internal field, spin wave interference in the vortex state, frequency splitting in rings with a tailored asymmetry, nonlinear spin dynamics and microwave assisted switching. Both micromagnetic simulations and semi-analytical calculations have been used to interpret the data and explain the microscopic origin of the reported phenomena. The detailed understanding of spin dynamics, the vortex circulation control and microwave assisted switching might be of technological interest if ring devices enter the field of magnonics.

Acknowledgments

The authors thank C. Back, M. Bolte, B. Botters, D. Görlitz, D. Heitmann, T. Korn, G. Meier, T. Rapp, P. Reuter, A. von Staa, M. Steiner, A. Wittmann for discussions and experimental support. We acknowledge A. Foncuberta i Morral for critical reading of the manuscript. We thank M. R. Freeman for the possibility to perform dynamic micromagnetic simulations at the University of Alberta. We thank the Deutsche Forschungsgemeinschaft for support via SFB 668 and in the framework of the German excellence initiative "Nanosystems Initiative Munich (NIM)".

References

1. Acremann, Y., Back, C. H., Buess, M., Portmann, O., Vaterlaus, A., Pescia, D. and Melchior, H. (2000), Imaging precessional motion of the magnetization vector, *Science* **290**, p. 492.

2. Agilent (1998), Network Analyzer Basics, *HP Back to Basics Seminar, Appliction note 95-1, www.home.agilent.com.*
3. Back, C. H., Allenspach, R., Weber, W., Parkin, S. S. P., Weller, D., Garwin, E. L. and Siegmann, H. C. (1999), Minimum field strength in precessional magnetization reversal, *Science* **285**, p. 864.
4. Back, C. H., Weller, D., Heidmann, J., Mauri, D., Guarisco, D., Garwin, E. L. and Siegmann, H. C. (1998), Magnetization reversal in ultrashort magnetic field pulses, *Phys. Rev. Lett.* **81**, p. 3251.
5. Bauer, M., Lopusnik, R., Fassbender, J. and Hillebrands, B. (2000), Suppression of magnetic-field pulse-induced magnetization precession by pulse tailoring, *Appl. Phys. Lett.* **76**, p. 2758.
6. Bayer, C., Demokritov, S. O., Hillebrands, B. and Slavin, A. N. (2003), Spin wave wells with multiple states created in small magnetic elements, *Appl. Phys. Lett.* **84**, p. 607.
7. Bayer, C., Jorzick, J., Hillebrands, B., Demokritov, S., Kouba, R., Bozinoski, R., Slavin, A. N., Guslienko, K., Berkov, D., Gorn, N. and Kostylev, M. P. (2005a), Spin wave excitations in finite rectangular elements of $Ni_{80}Fe_{20}$, *Phys. Rev. B* **72**, p. 064427.
8. Bayer, C., Park, J. P., Wang, H., Yan, M., Campbell, C. E. and Crowell, P. A. (2004), Spin waves in an inhomogeneously magnetized stripe, *Phys. Rev. B* **69**, p. 134401.
9. Bayer, C., Schultheiss, H., Hillebrands, B. and Stamps, R. L. (2005b), Phase shift of spin waves traveling through a 180 ° Bloch domain wall, *IEEE Trans. on Mag.* **41**, p. 3094.
10. Berkov, D. V. and Gorn, N. L. (2007), MicroMaguspackage for micromagnetic simulations, http://www.micromagus.de.
11. Bloembergen, N. and Wang, S. (1954), Relaxation effects in para- and ferromagnetic resonance, *Phys. Rev.* **93**, p. 72.
12. Botters, B., Giesen, F., Podbielski, J., Bach, P., Schmidt, G., Molenkamp, L. W. and Grundler, D. (2006), Stress dependence of ferromagnetic resonance and magnetic anisotropy in a thin NiMnSb film on InP(001), *Appl. Phys. Lett.* **89**, p. 242505.
13. Buess, M., Höllinger, R., Haug, T., Perzlmaier, K., Krey, U., Pescia, D., Scheinfein, M. R., Weiss, D. and Back, C. H. (2004), Fourier transform imaging of spin vortex eigenmodes, *Phys. Rev. Lett.* **93**, p. 077207.
14. Choi, S., Lee, K.-S., Guslienko, K. Y. and Kim, S.-K. (2007), Strong radiation of spin waves by core reversal of a magnetic vortex and their wave behaviors in magnetic nanowire waveguides, *Phys. Rev Lett.* **98**, p. 087205.
15. Covington, M., Crawford, T. M. and Parker, G. J. (2002), Time-resolved measurement of propagating spin waves in ferromagnetic thin films, *Phys. Rev. Lett.* **89**, p. 237202.
16. Crawford, T. M., Covington, M. and Parker, G. J. (2003), Time-domain excitation of quantized magnetostatic spin-wave modes in patterned NiFe thin film ensembles, *Phys. Rev. B* **67**, p. 024411.

17. CST (2006), Computer Simulation Technology, CST Microwave Studio 2006, www.cst.com.
18. Damon, R. W. (1953), Relaxation effects in the ferromagnetic resonance, *Rev. Mod. Phys.* **25**, p. 239.
19. Demidov, V., Demokritov, S., Hillebrands, B. and Laufenberg, M. (2004), Radiation of spin waves by a single micrometer-sized magnetic element, *Appl. Phys. Lett.* **85**, p. 2866.
20. Demokritov, S. O., Serga, A. A., André, A., Demidov, V. E., Kostylev, M. P., Hillebrands, B. and Slavin, A. N. (2004), Tunneling of dipolar spin waves through a region of inhomogeneous magnetic field, *Phys. Rev. Lett.* **93**, p. 047201.
21. Donahue, M. J. and Porter, D. G. (Sept. 1999), OOMMF User's Guide, *Interagency Report NISTIR6376, National Institute of Standards and Technology Gaithersburg, MD.*
22. Farle, M. (1998), Ferromagnetic resonance of ultrathin metallic layers, *Rep. Prog. Phys.* **61**, p. 755.
23. Freeman, M. R. and Hiebert, W. K. (2003), *Spin Dynamics in Confined Magnetic Structures I*, Topics in Applied Physics (Springer-Verlag, Berlin, Heidelberg, New York), pp. 93–126.
24. Gerrits, T., van den Berg, H. A. M., Hohlfeld, J., Bär, L. and Rasing, T. (2002), Ultrafast precessional magnetization reversal by picosecond magnetic-field pulse shaping, *Nature* **418**, p. 509.
25. Giesen, F. (2005), *Magnetization Dynamics of Nanostructured Ferromagnetic Rings and Rectangular Elements*, Ph.D. thesis, Institut für Angewandte Physik, Universität Hamburg, Germany.
26. Giesen, F., Podbielski, J., Botters, B. and Grundler, D. (2007a), Vortex circulation control in large arrays of asymmetric magnetic rings, *Phys. Rev. B* **75**, p. 184428.
27. Giesen, F., Podbielski, J. and Grundler, D. (2007b), Mode localization transition in ferromagnetic microscopic rings, *Phys. Rev. B* **76**, p. 014431.
28. Giesen, F., Podbielski, J., Korn, T. and Grundler, D. (2005a), Multiple ferromagnetic resonance in mesoscopic permalloy rings, *J. Appl. Phys.* **97**, p. 142501.
29. Giesen, F., Podbielski, J., Korn, T., Steiner, M., van Staa, A. and Grundler, D. (2005b), Hysteresis and control of ferromagnetic resonances in microscopic rings, *Appl. Phys. Lett.* **86**, p. 112510.
30. Griffiths, J. H. E. (1946), *Nature* **158**, p. 670.
31. Grundler, D. (2002), Spintronics, *Physics World* **April**, p. 39.
32. Grundler, D., Giesen, F. and Podbielski, J. (2007), Tailoring spin excitations by nanolithography, *Physics in Canada* **63**, p. 63.
33. Grundler, D., Hengstmann, T., Klockmann, N., Heyn, C. and Heitmann, D. (2002), Bend-resistance nanomagnetometry: spatially resolved magnetization studies in ferromagnet/semiconductor hybrid structures, *Physica E* **12**, p. 248.
34. Guslienko, K. Y., Chantrell, R. W. and Slavin, A. N. (2003), Dipolar localization of quantized spin-wave modes in thin rectangular magnetic elements,

Phys. Rev. B **68**, p. 024422.

35. Guslienko, K. Y., Demokritov, S. O., Hillebrands, B. and Slavin, A. N. (2002), Effective dipolar boundary conditions for dynamic magnetization in thin magnetic stripes, *Phys. Rev. B* **66**, p. 132402.
36. Guslienko, K. Y. and Slavin, A. N. (2005), Boundary conditions for magnetization in magnetic nanoelements, *Phys. Rev. B* **72**, 014463.
37. Heinrich, B. and Bland, J. A. C. (1994), *Ultrathin Magnetic Structures, Vols. I and II* (Springer),
38. Hengstmann, T., Grundler, D., Heyn, C. and Heitmann, D. (2001), Stray-field investigation on permalloy nanodiscs, *J. Appl. Phys.* **90**, p. 6542.
39. Hertel, R., Wulfhekel, W. and Kirschner, J. (2004), Domain-wall induced phase shifts in spin waves, *Phys. Rev Lett.* **93**, p. 257202.
40. Hiebert, W. K., Stankiewicz, A. and Freeman, M. R. (1997), Direct observation of magnetic relaxation in a small permalloy disk by time-resolved scanning Kerr microscopy, *Phys. Rev. Lett.* **79**, p. 1134.
41. Hillebrands, B. and Ounadjela, K. (2002), *Spin Dynamics in Confined Magnetic Structures I & II*, Topics in Applied Physics (Springer).
42. Hillebrands, B. and Thiaville, A. (2006), *Spin Dynamics in Confined Magnetic Structures III*, Topics in Applied Physics (Springer).
43. Jackson, J. D. (1999), *Classical Electrodynamics*, 3rd edn. (John Wiley & Sons, Inc.).
44. Jorzick, J., Demokritov, S. O., Hillebrands, B., Bailleul, M., Fermon, C., Guslienko, K. Y., Slavin, A. N., Berkov, D. V. and Gorn, N. L. (2002), Spin wave wells in nonellipsoidal micrometer size magnetic elements, *Phys. Rev. Lett.* **88**, p. 047204.
45. Jorzick, J., Demokritov, S. O., Mathieu, C., Hillebrands, B., Bartenlian, B., Chappert, C., Rousseaux, F. and Slavin, A. N. (1999), Brillouin light scattering from quantized spin waves in micron-size magnetic wires, *Phys. Rev. B* **60**, p. 15194.
46. Joseph, R. I. and Schlömann, E. (1965), Demagnetizing field in nonellipsoidal bodies, *J. Appl. Phys.* **36**, p. 1579.
47. Kalinikos, B. A. and Slavin, A. N. (1986), Theory of dipole-exchange spin wave spectrum for ferromagnetic films with mixed exchange boundary conditions, *J. Phys. C: Solid State Phys.* **19**, p. 7013.
48. Kittel, C. (1948), On the theory of ferromagnetic resonance absorption, *Phys. Rev.* **73**, p. 155.
49. Kittel, C. (1958), Excitation of spin waves in a ferromagnet by a uniform rf field, *Phys. Rev.* **110**, p. 1295.
50. Kläui, M., Rothman, J., Lopez-Diaz, L., Vaz, C. A. F. and Bland, J. A. C. (2001), Vortex circulation control in mesoscopic ring magnets, *Appl. Phys. Lett.* **78**, p. 3268.
51. Kläui, M., Vaz, C. A. F., Lopez-Diaz, L. and Bland, J. A. C. (2003), Vortex formation in narrow ferromagnetic rings, *J. Phys.: Condens. Matter* **15**, pp. R985–R1023.
52. Korn, T., Giesen, F., Podbielski, J., Ravlic, D., Schueller, C. and Grundler, D. (2005), Time-resolved study of the increased magnetization precession

frequency in Fe wires, *J. Magn. Magn. Mater* **285**, p. 240.

53. Landau, L. D. and Lifshitz, E. (1935), On the theory of the dispersion of magnetic permeability in ferromagnetic bodies, *Phys. Z. Sowjetunion* **8**, p. 153.
54. Last, T., Hacia, S., Fischer, S. F. and Kunze, U. (2006), Aging phenomena in single nanopatterned permalloy wires, *Physica B* **384**, pp. 9–11.
55. Li, S. P., Peyrade, D., Natali, M., Lebib, A., Chen, Y., Ebels, U., Buda, L. D. and Ounadjela, K. (2001), Flux closure structures in cobalt rings, *Phys. Rev. Lett.* **86**, p. 1102.
56. Mathieu, C., Jorzick, J., Frank, A., Demokritov, S. O., Slavin, A. N., Hillebrands, B., Bartenlian, B., Chappert, C., Decanini, D., Rousseaux, F. and Cambril, E. (1998), Lateral quantization of spin waves in micron-size magnetic wires, *Phys. Rev. Lett.* **81**, p. 3968.
57. Meier, G., Grundler, D., Broocks, K.-B., Heyn, C. and Heitmann, D. (2000), Effect of tilted magnetic fields on bistable nanomagnets in hybrid semiconductor/ferromagnet devices, *J. Magn. Magn. Mater.* **210**, p. 138.
58. Neudecker, I., Kläui, M., Perzlmaier, K., Backes, D., Heydermann, L. J., Vaz, C. A. F., Bland, J. A. C., Rüdiger, U. and Back, C. H. (2006a), Spatially resolved dynamic eigenmode spectrum of Co rings, *Phys. Rev. Lett.* **96**, p. 057207.
59. Neudecker, I., Woltersdorf, G., Heinrich, B., Okuno, T., Gubbiotti, G. and Back, C. H. (2006b), Comparison of frequency, field, and time domain ferromagnetic resonance methods, *J. Magn. Magn. Mater.* **307**, p. 148.
60. Park, J. P. and Crowell, P. A. (2005), Interactions of spinwaves with a magnetic vortex, *Phys. Rev. Lett.* **95**, p. 167201.
61. Park, J. P., Eames, P., Engebretson, D. M., Berezovsky, J. and Crowell, P. A. (2002), Spatially resolved dynamics of localized spin-wave modes in ferromagnetic wires, *Phys. Rev. Lett.* **89**, p. 277201.
62. Podbielski, J., Giesen, F., Berginski, M., Hoyer, N. and Grundler, D. (2005), Spin configurations in nanostructured magnetic rings: from DC transport to GHz spectroscopy, *Superlattices and Microstructures* **37**, p. 341.
63. Podbielski, J., Giesen, F. and Grundler, D. (2006), Spin-wave interference in microscopic rings, *Phys. Rev. Lett.* **96**, p. 167207.
64. Podbielski, J., Heitmann, D. and Grundler, D. (2007), Microwave assisted switching of microscopic rings: Correlation between nonlinear spin dynamics and critical microwave fields, *Phys. Rev. Lett.* **99**, p. 207202.
65. Rolff, H., Pfützner, W., Heyn, C. and Grundler, D. (2004), Hall magnetometry on a ferromagnetic nanoring, *Journal of Magnetism and Magnetic Materials* **272–276**, p. 1623.
66. Rothman, J., Kläui, M., Lopez-Diaz, L., Vaz, C. A. F., Bleloch, A., Bland, J. A. C., Cui, Z. and Speaks, R. (2001), Observation of a bi-domain state and nucleation free switching in mesoscopic ring magnets, *Phys. Rev. Lett.* **86**, p. 1098.
67. Scheinfein, M. R. (1997), LLG Micromagnetics Simulator, http://llgmicro.-home.mindspring.com.

68. Schumacher, H. W., Chappert, C., Crozat, P., Sousa, R. C., Freitas, P. P., Miltat, J., Fassbender, J. and Hillebrands, B. (2003a), Phase coherent magnetization reversal in microscopic spin valve elements, *Phys. Rev. Lett.* **90**, p. 017201.
69. Schumacher, H. W., Chappert, C., Sousa, R. C., Freitas, P. P. and Miltat, J. (2003b), Quasiballistic magnetization reversal, *Phys. Rev. Lett.* **90**, p. 017204.
70. Silva, T. J., Lee, C. S., Crawford, T. M. and Rogers, C. T. (1999), Inductive measurements of ultrafast magnetization dynamics in thin-film Permalloy, *J. Appl. Phys.* **85**, p. 7849.
71. SonnetLite (2007), http://www.sonnetusa.com/products/lite/.
72. Suhl, H. (1957), The theory of ferromagnetic resonance at high signal powers, *J. Phys. Chem. Solids.* **1**, p. 209.
73. Thirion, C., Wernsdorfer, W. and Mailly, D. (2005), Switching of magnetization by nonlinear resonance studied in single nanoparticles, *Appl. Phys. Lett.* **86**, p. 262502.
74. Tudosa, I., Stamm, C., Kashuba, A. B., King, F., Siegmann, H. C., Stöhr, J., Ju, G., Lu, B. and Weller, D. (2004), The ultimate speed of switching in granular recording media, *Nature* **428**, p. 831.
75. Viau, G., Fiévet-Vincent, F., Fiévet, F., Toneguzzo, P., Ravel, F. and Acher, O. (1997), Size dependence of microwave permeability of spherical ferromagnetic particles, *J. Appl. Phys* **81**, p. 2749.
76. Yager, W. A. and Bozorth, R. M. (1947), Ferromagnetic resonance at microwave frequencies, *Phys. Rev. B* **72**, p. 80.
77. Zhu, X., Liu, Z., Metlushko, V., Grutter, P. and Freeman, M. R. (2005a), Broadband spin dynamics of the magnetic vortex state: Effect of the pulsed field direction, *Phys. Rev. B* **71**, p. 180408(R).
78. Zhu, X., Malac, M., Liu, Z., Qian, H., Metlushko, V. and Freeman, M. R. (2005b), Broadband spin dynamics of Permalloy rings in the circulation state, *Appl. Phys. Lett.* **86**, p. 262502.

Chapter 6

LOCALIZED AND PROPAGATING SPIN WAVE MODES EXCITED IN MAGNETIC NANO-STRUCTURES BY SPIN-POLARIZED CURRENT

Andrei Slavin and Vasil Tiberkevich

Department of Physics, Oakland University, Rochester, MI 48309, USA

1. Introduction

It was discovered theoretically by Slonczhewski and Berger[1–3] that spin polarized electric current passing through a magnetized ferromagnetic layer can transfer its angular momentum to this layer, and, as a result, change the orientation of the magnetization vector in this layer. This phenomenon — named spin-torque effect — can lead either to the magnetization switching (reversal of the magnetization direction), or to the magnetization precession with the frequency close to the frequency of the ferromagnetic resonance (FMR) in the magnetic layer. Since the FMR frequencies in thin magnetic films typically lie in the microwave range, the spin-torque effect creates a novel physical mechanism for the generation of microwave signals in nano-sized magnetic systems.

A clear theoretical understanding of the process of excitations of microwave oscillations of magnetization by spin-polarized current is necessary both for the further development of fundamental physics of nano-magnetic systems and for the development of practical nano-sized active microwave devices based on the spin-torque effect — spin-torque nano-oscillators.

In contrast with many the other methods of excitation of magnetization precession in magnetically ordered substances(such as linear FMR or parametric parallel pumping) spin-torque effect is very efficient, and allows us to achieve very large precession angles reaching tens of degrees for bias currents that exceed the excitation threshold by not more than 50 percent. When the precession angles are so large the nonlinearity of the spin system plays a very important role in the magnetization dynamics caused by the spin-torque effect. In particular, this nonlinearity can lead to a *new type of spin wave confinement*, when a spin wave mode excited by spin-torque effect becomes evanescent, and , therefore, non-propagating and confined, due to the negative nonlinear shift of its frequency. We discuss below this type of nonlinear spin wave confinement in detail, and demonstrate that a self-localized two-dimensional non-propagating spin wave packet — standing spin wave "bullet" — appearing in the presence of dissipation and competing with it spin-torque effect, is stable and has the lowest excitation threshold. Thus, the self-localized and spatially confined spin wave "bullet" is the spin wave mode that is excited in real spin-torque experiments in magnetic nano-contacts.

In this Chapter we develop analytic theory of current-induced microwave excitations in magnetic films based on the classical quasi-Hamiltonian formalism for spin waves.[4] First of all, we present the Landau–Lifshits equation of motion for the variable magnetization of a magnetic film ("free" layer of a magnetic multi-layered structure described in[1,3] where dissipative torque and Slonczewski's spin-torque are taken into account and are treated as small non-conservative perturbations compared to the main conservative "precessional" torque caused by the total effective magnetic field $\boldsymbol{H}_{\mathrm{eff}}$.

We show that for a certain direction of the electric current the spin-torque has a component that creates effective negative damping and, for sufficiently large current densities, can completely compensate natural positive damping and lead to the instability of spin wave modes of the layer. We also demonstrate that both dissipative torque and spin-torque are nonlinear (i.e. strongly dependent on the precession angle of the excited mode), and that the competition between these nonlinear torques determines the equilibrium amplitude of the microwave oscillation excited in the magnetic layer by the spin-polarized current. One of the most important conclusions of our analytic theory is the statement that the current-induced magnetization dynamics of a thin magnetic layer can be successfully described by a simple nonlinear equation equivalent to the equation describing dynamics of a nonlinear auto-oscillator.

Then, we take into account the spatial distribution of the current density in the experimental sample (section 4.2) and determine the character and spatial distribution of the spin wave modes excites by spin-polarized electric current in a magnetic layer. We demonstrate that both quasi-linear propagating and nonlinear self-localized spin wave modes could be excited and show that in the case of an in-plane magnetized magnetic layer the lowest excitation threshold corresponds to the self-localized nonlinear "bullet" mode.

Although the majority of our results in this Chapter are obtained for the case of a so-called magnetic nano-contacts (i.e. in the geometry where the free magnetic layer in which microwave oscillations of magnetization take place is infinitely extended in-plane), the developed formalism can be easily extended for the case of a magnetic nano-pillar (i.e the geometry where the free layer has finite in-plane dimensions). In the latter case the spectrum of spin wave modes becomes discrete due to the frequency quantization caused by the finite in-plane sizes of the free magnetic layer, and the boundary conditions for the variable magnetization at the lateral edges of the magnetic free layer (see Section 2.4) should be taken into account.

2. Landau–Lifshits–Gilbert–Slonczewski Equation of Magnetization Dynamics

Dynamics of the magnetization $\boldsymbol{M} = \boldsymbol{M}(t, \boldsymbol{r})$ of the "free" ferromagnetic layer in the presence of spin-polarized current is described by the Landau–Lifshits–Gilbert equation with an additional Slonczewski–Berger spin-transfer term:[1–3]

$$\frac{\partial \boldsymbol{M}}{\partial t} = \gamma \left[\boldsymbol{H}_{\mathrm{eff}} \times \boldsymbol{M} \right] + \boldsymbol{T}_{\mathrm{d}} + \boldsymbol{T}_{\mathrm{s}} \, . \tag{1}$$

The first term in the RHS of this equation describes conservative precession of the magnetization vector $\boldsymbol{M}$ about the direction of the effective magnetic field $\boldsymbol{H}_{\mathrm{eff}}$ ($\gamma = g\mu_{\mathrm{B}}/\hbar \approx 2.8$ MHz/Oe is the modulus of the gyromagnetic ratio, g is the spectroscopic Lande factor, μ_{B} is the Bohr magneton, and $\hbar$ is the reduced Planck costant), the second term $\boldsymbol{T}_{\mathrm{d}}$ is the dissipative torque describing energy dissipation, and the last term $\boldsymbol{T}_{\mathrm{s}}$ is the Slonczewski-Berger spin-transfer torque describing interaction of the magnetization $\boldsymbol{M}$ with the spin-polarized current traversing the "free" ferromagnetic layer.

2.1. *Effective magnetic field* $\boldsymbol{H}_{\text{eff}}$

The effective magnetic field $\boldsymbol{H}_{\text{eff}}$ can be calculated as a variational derivative of the free energy W of the ferromagnetic layer with respect to magnetization vector:[5]

$$\boldsymbol{H}_{\text{eff}} = -\frac{\delta W}{\delta \boldsymbol{M}} . \tag{2}$$

There are several main contributions into the free energy W: Zeeman energy of interaction with external magnetic field, magnetodipolar energy, energy of crystallographic anisotropy, energy of inhomogeneous intralayer exchange, and energy of interaction with the Oersted magnetic field created by the current. Respectively, effective field $\boldsymbol{H}_{\text{eff}}$ also can be written as a sum of five terms:

$$\boldsymbol{H}_{\text{eff}} = \boldsymbol{H}_{\text{ext}} + \boldsymbol{H}_{\text{dip}} + \boldsymbol{H}_{\text{an}} + \boldsymbol{H}_{\text{ex}} + \boldsymbol{h}_I . \tag{3}$$

2.1.1. *External magnetic field* $\boldsymbol{H}_{\text{ext}}$

In general case the external magnetic field $\boldsymbol{H}_{\text{ext}}$ depends both on spatial coordinates $\boldsymbol{r}$ and time t. Due to small nano-scale sizes of the considered magnetic system, however, one can safely assume that $\boldsymbol{H}_{\text{ext}}$ is spatially-uniform. Concerning the time dependence of $\boldsymbol{H}_{\text{ext}}$, we will consider the case when the external field is a sum of constant bias magnetic field $\boldsymbol{H}_0$ and microwave magnetic field $\boldsymbol{h}_\sim = \boldsymbol{h}_\sim(t)$:

$$\boldsymbol{H}_{\text{ext}} = \boldsymbol{H}_0 + \boldsymbol{h}_\sim . \tag{4}$$

The variable magnetic field $\boldsymbol{h}_\sim$ describes direct[5] or parametric[4] excitation of spin waves by magnetic field and, also, it provides external phase locking signal for the current-driven microwave precession.[6] Due to the smallness of real microwave magnetic fields compared to the bias magnetic field it is natural to consider $\boldsymbol{h}_\sim$ as a small perturbation.

Note that the small thickness of metallic spacer between the “free” and “fixed” magnetic layers causes significant RKKY interlayer exchange interaction between the ferromagnetic layers. This effect can be easily taken into account by renormalization of the bias magnetic field $\boldsymbol{H}_0 \Rightarrow \boldsymbol{H}_{\text{app}} + \boldsymbol{H}_{\text{RKKY}}$, where $\boldsymbol{H}_{\text{app}}$ is the applied magnetic field (i.e. "real" external field) and $\boldsymbol{H}_{\text{RKKY}}$ is the effective field of the RKKY interlayer interaction. The magnetic dipolar stray field created by the “fixed” magnetic layer will lead to the analogous renormalization of $\boldsymbol{H}_0$.

In coordinate form bias magnetic field $\boldsymbol{H}_0$ can be written as

$$\boldsymbol{H}_0 = (\sin\theta_0\,\widehat{\mathbf{z}} + \cos\theta_0\cos\phi_0\,\widehat{\mathbf{x}} + \cos\theta_0\sin\phi_0\,\widehat{\mathbf{y}})H_0\,, \tag{5}$$

where H_0, θ_0, and ϕ_0 are the magnitude, *out-of-plane* angle, and angle with the axis $\widehat{\mathbf{x}}$ of the bias magnetic field, respectively.

2.1.2. *Magnetodipolar field* $\boldsymbol{H}_{\mathrm{dip}}$

Magnetodipolar field $\boldsymbol{H}_{\mathrm{dip}}$ in Eq. (2) is determined from the Maxwell equations in magnetostatic approximation:[5]

$$\operatorname{curl}\boldsymbol{H}_{\mathrm{dip}} = \mathbf{0}\,, \tag{6a}$$

$$\operatorname{div}\boldsymbol{H}_{\mathrm{dip}} = -4\pi\operatorname{div}\boldsymbol{M}\,. \tag{6b}$$

For the considered thin-film geometry solution for $\boldsymbol{H}_{\mathrm{dip}}$ can be written in the form of Fourier series:

$$\boldsymbol{H}_{\mathrm{dip}}(\boldsymbol{r},t) = -4\pi\sum_{\boldsymbol{k}}\widehat{\boldsymbol{G}}_{\boldsymbol{k}}\cdot\boldsymbol{M}_{\boldsymbol{k}}(t)e^{i\boldsymbol{k}\cdot\boldsymbol{\rho}}\,, \tag{7}$$

where $\boldsymbol{k} = k_x\widehat{\mathbf{x}} + k_y\widehat{\mathbf{y}}$ is the two-dimensional in-plane wavevector and $\widehat{\boldsymbol{G}}_{\boldsymbol{k}}$ is the thin-film magnetostatic tensor Green's function:

$$\widehat{\boldsymbol{G}}_{\boldsymbol{k}} = \begin{pmatrix} (1-P_k)\dfrac{k_x^2}{k^2} & (1-P_k)\dfrac{k_xk_y}{k^2} & 0 \\ (1-P_k)\dfrac{k_xk_y}{k^2} & (1-P_k)\dfrac{k_y^2}{k^2} & 0 \\ 0 & 0 & P_k \end{pmatrix}. \tag{8}$$

Here P_k is given by

$$P_k = \frac{1-e^{-kd}}{kd}\,, \tag{9}$$

where d is the thickness of the magnetic film.

In Eq. (7) $\boldsymbol{\rho} = x\,\widehat{\mathbf{x}} + y\,\widehat{\mathbf{y}}$ is the two-dimensional radius vector in the film plane and $\boldsymbol{M}_{\boldsymbol{k}}(t)$ is the Fourier harmonics of the magnetization with wavevector $\boldsymbol{k}$:

$$\boldsymbol{M}(\boldsymbol{r},t) = \sum_{\boldsymbol{k}}\boldsymbol{M}_{\boldsymbol{k}}(t)e^{i\boldsymbol{k}\cdot\boldsymbol{\rho}}\,. \tag{10}$$

Here we have assumed that the magnetization $\boldsymbol{M}$ is uniform across the film thickness (i.e. $\boldsymbol{M}$ does not depend on the normal-to-the-film coordinate z), which can be done for the lowest-frequency magnetic excitations of a thin ferromagnetic film without substantial surface anisotropy.

Note that for thin ($d \to 0$) magnetic film $P_k \to 1$ and, respectively, Green's tensor become $\boldsymbol{k}$-independent, $\widehat{\boldsymbol{G}}_{\boldsymbol{k}} \approx \widehat{\mathbf{z}}\widehat{\mathbf{z}}$. In this *thin-film approximation* magnetodipolar field takes very simple form

$$\boldsymbol{H}_{\rm dip} = -4\pi(\boldsymbol{M}\cdot\widehat{\mathbf{z}})\widehat{\mathbf{z}}\ . \qquad (11)$$

Thin-film approximation can be used when the film thickness d is much smaller than the characteristic in-plane sizes of the magnetic system (e.g., radius of the nano-contact or nano-pillar) and when it is smaller or of the order of the exchange length $\lambda_{\rm ex}$ of the ferromagnetic material. The typical spin-torque experiments are performed exactly in this range of parameters, so the expression Eq. (11) gives a rather good approximation of the dipolar magnetic field and will be used below instead of the exact Eq. (7).

2.1.3. *Effective field of crystallographic anisotropy* $\boldsymbol{H}_{\rm an}$

We will consider the most common case of an easy-axis magnetic crystallographic anisotropy, and assume that the anisotropy axis lies in the film plane along the direction $\widehat{\mathbf{x}}$. In this case effective anisotropy field $\boldsymbol{H}_{\rm an}$ can be written as[5]

$$\boldsymbol{H}_{\rm an} = \frac{H_a}{M_0}(\boldsymbol{M}\cdot\widehat{\mathbf{x}})\,\widehat{\mathbf{x}}\,, \qquad (12)$$

where $H_a > 0$ is the anisotropy field (case $H_a < 0$ corresponds to the hard-axis anisotropy) and $M_0 = |\boldsymbol{M}|$ is the saturation magnetization. Anisotropy field H_a is connected with the crystallographic anisotropy energy constant K_a by $H_a = 2K_a/M_0$.

2.1.4. *Effective field of inhomogeneous exchange* $\boldsymbol{H}_{\rm ex}$

Effective field of the inhomogeneous exchange $\boldsymbol{H}_{\rm ex}$ can be written as[5]

$$\boldsymbol{H}_{\rm ex} = 4\pi\lambda_{\rm ex}^2\nabla^2\boldsymbol{M}\,, \qquad (13)$$

where $\lambda_{\rm ex} = \sqrt{A/2\pi M_0^2}$ is the exchange length (A is the exchange stiffness), and ∇^2 is the Laplace operator. For most magnetic metals $\lambda_{\rm ex} \approx 5$ nm. Taking into account that the magnetization $\boldsymbol{M}$ can be considered to be uniform across the film thickness, we can assume that the Laplace operator acts only on the in-plane coordinates, i.e.

$$\nabla^2 = \frac{\partial^2}{\partial x^2} + \frac{\partial^2}{\partial y^2}\ . \qquad (14)$$

2.1.5. *Oersted field $\boldsymbol{h}_I$ created by the current*

Electric current traversing the "free" magnetic layer creates its own Oersted magnetic field $\boldsymbol{h}_I$, which also interacts with the magnetization. Oersted magnetic field $\boldsymbol{h}_I$ is determined from the static Maxwell equations

$$\operatorname{curl} \boldsymbol{h}_I = \frac{4\pi}{c}\boldsymbol{j}\,, \tag{15a}$$

$$\operatorname{div} \boldsymbol{h}_I = 0\,, \tag{15b}$$

where c is the speed of light and $\boldsymbol{j} = \boldsymbol{j}(t, \boldsymbol{r})$ is the density of the electric current.

Precise determination of the Oersted field $\boldsymbol{h}_I$ requires knowledge of the current distribution not only within the "free" magnetic layer, but also in all outer space. Here we will use the simplest model of current distribution, assuming that the current is uniform in $\widehat{\mathbf{z}}$ direction, i.e.

$$\boldsymbol{j}(t, \boldsymbol{r}) = j(t, \boldsymbol{\rho})\widehat{\mathbf{z}}\,, \tag{16}$$

where $j(t, \boldsymbol{\rho})$ is the (scalar) current density within the "free" layer. In this case $\widehat{\mathbf{z}}$ component of the Oersted field vanishes, and $\boldsymbol{h}_I$ can be written in the form

$$\boldsymbol{h}_I = \frac{\partial A_I}{\partial y}\widehat{\mathbf{x}} - \frac{\partial A_I}{\partial x}\widehat{\mathbf{y}}\,, \tag{17}$$

where A_I is the $\widehat{\mathbf{z}}$-component of the vector-potential, determined from the equation

$$\nabla^2 A_I = -\frac{4\pi}{c} j\,. \tag{18}$$

In the important model case of a circular nano-contact (or nano-pillar) of the radius R with uniform current distribution Oersted field is aligned along the direction $\widehat{\boldsymbol{\phi}}$ of the polar coordinate system and has the form

$$\boldsymbol{h}_I = \widehat{\boldsymbol{\phi}}\,\frac{2I}{c} \times \begin{cases} \dfrac{\rho}{R^2}\,, \ \rho < R \\[2ex] \dfrac{1}{\rho}\,, \quad \rho > R \end{cases}\,, \tag{19}$$

where $I = \pi R^2 j$ is the total current through the nano-contact.

Oersted magnetic field $\boldsymbol{h}_I$ can depend on time through the time-dependence of the bias current, and, also, it strongly depends on the coordinate. This is the main qualitative difference between the current-induced Oersted field $\boldsymbol{h}_I$ and external magnetic field $\boldsymbol{H}_{\text{ext}}$. In particular, this means that the Oersted field $\boldsymbol{h}_I$ can directly excite non-uniform spin wave modes, which are not coupled to the uniform external field.

2.2. *Gilbert dissipative torque* $\boldsymbol{T}_\mathrm{d}$

The dissipative processes in a magnetic system can be phenomenologically accounted for by an additional dissipative torque $\boldsymbol{T}_\mathrm{d}$ in the equation of magnetization dynamics (1). In the case of ferromagnetic metals an adequate description of the dissipation is given by the Gilbert-like damping torque[7,8]

$$\boldsymbol{T}_\mathrm{d} = \frac{\alpha(\xi)}{M_0}\left[\boldsymbol{M} \times \frac{\partial \boldsymbol{M}}{\partial t}\right], \tag{20}$$

where $\alpha(\xi)$ is a positive-definite dimensionless function of the dimensionless parameter

$$\xi = \frac{\left|\partial \boldsymbol{M}/\partial t\right|^2}{\omega_M^2 M_0^2}, \tag{21}$$

where $\omega_M = 4\pi\gamma M_0$. The parameter ξ is the dimensionless measure of the power of the magnetization precession, $\xi \sim |\boldsymbol{m}_\sim|^2/M_0^2$, where $\boldsymbol{m}_\sim$ is alternating part of the magnetization. Eq. (20) is the most general phenomenological form of the dissipative torque under the assumptions of isotropy, locality, conservation of the saturation magnetization ($|\boldsymbol{M}| = M_0 = \mathrm{const}$), and dynamical origin (i.e. dissipation is determined only by the local and instant values of $\boldsymbol{M}$ and $\partial\boldsymbol{M}/\partial t$) of the dissipative processes.[8]

In the framework of the phenomenological approach one can not reconstruct the function $\alpha(\xi)$. Nevertheless, for finite, but not very large angles of magnetization precession one can consider that $\xi \ll 1$ and represent $\alpha(\xi)$ as a Taylor series expansion,

$$\alpha(\xi) \simeq \alpha_\mathrm{G}(1 + q_1\xi + q_2\xi^2 + \dots), \tag{22}$$

where $\alpha_\mathrm{G} \ll 1$ is the linear Gilbert damping parameter, and $q_i \sim 1$ are the phenomenological parameters characterizing nonlinear properties of damping processes.

It should be noted that for the qualitative (and even quantitative) description of "traditional" magnetic systems it is usually enough to account only for the zero-order term in the expansion Eq. (22), i.e. to use linear Gilbert model $\alpha(\xi) = \alpha_\mathrm{G}$. In the considered case of current-driven magnetic nano-structures, however, there are two circumstances that force us to use more precise model Eq. (22) with at least non-vanishing first-order term $q_1 \neq 0$. First, linear dissipation (zero-order term in Eq. (22)) is fully compensated by the effective current-induced negative damping (see

Sec. 4) and, therefore, can not provide a mechanism for limiting amplitudes of excited spin waves. Second, due to discretization of spin wave spectrum in nano-sized structures, nonlinear magnon-magnon interactions are effectively prohibited in such structures and can not limit spin wave amplitudes as well. Therefore, one has to account for higher-order terms in Eq. (22) even for qualitative description of experimental results.

2.3. *Slonczewski spin-transfer torque* $\boldsymbol{T}_\mathrm{s}$

The last term in Eq. (1) in the Slonczewski-Berger spin torque $\boldsymbol{T}_\mathrm{s}$ describing transfer of angular momentum (or magnetic moment) from the spin-polarized current to the magnetization of the "free" magnetic layer. This term can be written as[1,3]

$$\boldsymbol{T}_\mathrm{s} = \frac{\varepsilon\eta j(t,\boldsymbol{\rho})}{M_0}\left[\boldsymbol{M}\times[\boldsymbol{M}\times\widehat{\mathbf{p}}]\right], \tag{23}$$

where ε is the dimenionless spin-polarization efficiency ($|\varepsilon| \leq 1$), describing level of polarization of the current, $\eta = g\mu_\mathrm{B}/2eM_0 d$ (g is the spectroscopic Lande factor, μ_B is the Bohr magneton, e is the modulus of the electron charge, and d is the thickness of the "free" magnetic layer), and $\widehat{\mathbf{p}}$ is the unit vector in the direction of spin polarization of the current ($\widehat{\mathbf{p}}$ coincides with the direction of magnetization of the "fixed" magnetic layer).

Positive values of the current density j correspond to the current flowing from the "fixed" magnetic layer to the "free" one (note, that, since electron bears *negative* charge, the direction of motion of conducting electrons is opposite to the direction of the current). As it will be shown below (see Sec. 4), this direction of the current corresponds to instability of the parallel orientations of the "free" and "fixed" magnetic layers.

Spin-polarization efficiency ε, in principle, depends on the mutual orientation of the magnetizations of the "fixed" and "free" magnetic layers, i.e. $\varepsilon = \varepsilon(\widehat{\mathbf{p}}\cdot\boldsymbol{M}/M_0)$. In most cases, however, this dependence does not lead to any qualitative or significant quantitative effects. The only difference is the case when equilibrium orientations of magnetization of the "free" and "fixed" layers are either exactly parallel or exactly orthogonal to each other. Even in this case, however, all effects of the angular dependence of the polarization efficiency ε can be modelled by introducing formal "misalignment angle" between magnetizations of the layers. Therefore, in the following we will assume that $\varepsilon = \text{const}$.

Product $\varepsilon\eta j$ has the dimension of frequency, and, as it will be shown in Sec. 4, represents the characteristic growth rate of current-induced magnetization instability.

2.4. *Boundary conditions for the magnetization*

Equation (1) should be solved together with appropriate boundary conditions for the magnetization $\boldsymbol{M}$ at the edges of the "free" ferromagnetic layer. These boundary conditions are determined by the surface anisotropy of the ferromagnetic material and by dipole-exchange edge effects. For most ferromagnetic metals, however, surface anisotropy is negligible, and boundary conditions for the magnetization at the top and bottom edges of the "free" ferromagnetic layer is of the "unpinned" type, $\partial \boldsymbol{M}/\partial z = 0$. This allows one to assume that the magnetization is uniform across the thickness of the film — the assumption that was already used for the calculation of the dipolar magnetic field Eq. (7).

The boundary conditions for the magnetization at the lateral edges of the "free" ferromagnetic layer (in the case of magnetic nano-pillar) are determined mostly by the dipole-exchange edge effects. For the circular nano-pillar of the radius R these effective boundary conditions can be written in the form:[9,10]

$$\Lambda \frac{\partial \boldsymbol{M}}{\partial r} + \boldsymbol{M} \bigg|_{r=R} = 0\,, \tag{24}$$

where

$$\Lambda = \frac{d\left[2(6\ln 2 - 1) + 4\ln(R/d) + (\lambda_{\mathrm{ex}}/d)^2\right]}{2\pi\left[1 - \dfrac{K_s}{\pi d M_0^2}\right]}\,, \tag{25}$$

K_s is the surface anisotropy constant, and M_0 is the saturation magnetization of the "free" layer.

In the case when the thickness of the "free" ferromagnetic layer is substantially larger than the exchange length λ_{ex} in the material of the layer (but still much smaller than the layer radius R), the dipolar edge effects play the dominant role and create conditions of strong pinning $\Lambda/R \gg 1$ at the lateral edges of the "free" layer.

In the opposite case when the thickness d of the "free" layer is smaller than the exchange length λ_{ex} (and if the surface anisotropy K_s is negligible) the dipole-exchange edge effects are not important and one can use the "unpinned" boundary conditions at the lateral edges of the "free" layer:

$$\frac{\partial \boldsymbol{M}}{\partial r} = 0 \, . \tag{26}$$

In the case of a magnetic nano-contact, having the "free" layer that is unbound in plane, which we shall mainly consider in this Chapter, the boundary conditions (24) at the lateral edges of the layer are not important.

In contrast, in the case of magnetic nano-pillar, having "free" layer of a finite lateral size (see e.g. Ref. 11), the boundary conditions (24) strongly influence the spectrum of spin wave modes that can be excited by spin-polarized current. Therefore, to determine the discrete spectrum of possible spin wave modes in the case of a nano-pillar one has to solve the full boundary problem, which is similar to the spectral problems solved in Refs. 12, 13.

3. Derivation of Quasi-Hamiltonian Equations of Motion

In typical experimental cases dissipative $\boldsymbol{T}_{\rm d}$ and spin-transfer $\boldsymbol{T}_{\rm s}$ torques, as well as microwave $\boldsymbol{h}_\sim$ and Oersted $\boldsymbol{h}_I$ magnetic fields are much smaller than other terms in Eq. (1) (namely, bias $\boldsymbol{H}_0$, dipolar $\boldsymbol{H}_{\rm dip}$, anisotropy $\boldsymbol{H}_{\rm an}$, and exchange $\boldsymbol{H}_{\rm ex}$ magnetic fields), and, therefore, it is natural to consider them as small perturbations. Thus, one can rewrite Eq. (1) in the form

$$\frac{\partial \boldsymbol{M}}{\partial t} = \gamma \left[\boldsymbol{M} \times \frac{\delta W_0}{\delta \boldsymbol{M}} \right] + \boldsymbol{T}_\epsilon \, , \tag{27}$$

where W_0 is the "zero-order" free energy of the system:

$$\begin{aligned} W_0 &= -\int \left(\boldsymbol{H}_0 + \frac{1}{2}\boldsymbol{H}_{\rm dip} + \frac{1}{2}\boldsymbol{H}_{\rm an} + \frac{1}{2}\boldsymbol{H}_{\rm ex} \right) \cdot \boldsymbol{M} d\boldsymbol{r} \qquad (28) \\ &= \int \left(-\boldsymbol{H}_0 \cdot \boldsymbol{M} + 2\pi (\boldsymbol{M} \cdot \widehat{\mathbf{z}})^2 - \frac{H_a}{2M_0} (\boldsymbol{M} \cdot \widehat{\mathbf{x}})^2 + 2\pi \lambda_{\rm ex}^2 |\boldsymbol{\nabla} \boldsymbol{M}|^2 \right) d\boldsymbol{r} \, , \end{aligned}$$

and $\boldsymbol{T}_\epsilon$ is the small "perturbation" torque:

$$\boldsymbol{T}_\epsilon = \gamma \left[\widetilde{\boldsymbol{h}} \times \boldsymbol{M} \right] + \boldsymbol{T}_{\rm d} + \boldsymbol{T}_{\rm s} \, , \tag{29}$$

where $\widetilde{\boldsymbol{h}}$ is the sum of external microwave and current-induced Oersted magnetic fields:

$$\widetilde{\boldsymbol{h}} = \boldsymbol{h}_\sim + \boldsymbol{h}_I \, . \tag{30}$$

Due to the smallness of the perturbation terms one can use in dissipative torque (20) zero-order value for the time derivative of the magnetization, $\partial \boldsymbol{M}/\partial t \to \gamma[\boldsymbol{M} \times \delta W_0/\delta \boldsymbol{M}]$.

The most simple and physically transparent description of the dynamics of any perturbed system is achived in terms of "normal modes" of the unperturbed system. We shall proceed in this way, i.e. we shall find normal modes of the conservative magnetic system with free energy W_0 and then we shall find corrections to the dynamic equations for normal modes caused by the perturbation torque $\boldsymbol{T}_\epsilon$.

3.1. *Equilibrium orientation of the magnetization*

To find the normal modes of the unperturbed magnetic system, one has to find "ground state" of the system, i.e. stationary equilibrium magnetization $\boldsymbol{M}_{\text{eq}}$. With the unpinned boundary conditions Eq. (26) equilibrium magnetization will be spatially-uniform and one has to find only the equilibrium orientation $\widehat{\boldsymbol{\zeta}} = \boldsymbol{M}_{\text{eq}}/M_0$ of the magnetization. It follows form the Landau–Lifshits equation that for the equilibrium orientation zero-order effective magnetic field $-\delta W_0/\delta \boldsymbol{M}$ should be parallel to equilibrium magnetization orientation $\widehat{\boldsymbol{\zeta}}$ and, therefore, can be written as

$$-\frac{\delta W_0}{\delta \boldsymbol{M}} = H\widehat{\boldsymbol{\zeta}}, \tag{31}$$

where H is the *internal* magnetic field. Using Eq. (28) one can write the above condition explicitly:

$$\boldsymbol{H}_0 - 4\pi M_0(\widehat{\boldsymbol{\zeta}} \cdot \widehat{\mathbf{z}})\widehat{\mathbf{z}} + H_a(\widehat{\boldsymbol{\zeta}} \cdot \widehat{\mathbf{x}})\widehat{\mathbf{x}} = H\widehat{\boldsymbol{\zeta}}\ . \tag{32}$$

This vector equation allows one to find the magnitude H and direction $\widehat{\boldsymbol{\zeta}}$ of the internal magnetic field.

Using coordinate representation of the equilibrium orientation $\widehat{\boldsymbol{\zeta}}$,

$$\widehat{\boldsymbol{\zeta}} = \sin\theta\,\widehat{\mathbf{z}} + \cos\theta\cos\phi\,\widehat{\mathbf{x}} + \cos\theta\sin\phi\,\widehat{\mathbf{y}}\,, \tag{33}$$

where θ and ϕ are *internal* magnetization angles, one can rewrite Eq. (32) in the coordinate form as

$$H\sin\theta = H_0\sin\theta_0 - 4\pi M_0\sin\theta\,, \tag{34a}$$

$$H\cos\theta\cos\phi = H_0\cos\theta_0\cos\phi_0 + H_a\cos\theta\cos\phi\,, \tag{34b}$$

$$H\cos\theta\sin\phi = H_0\cos\theta_0\sin\phi_0\ . \tag{34c}$$

In general case Eq. (34) can be reduced to 6-order equation for H and, therefore, can be solved only numerically. Note that for $H_a = 0$, when H represents real magnetic field inside ferromagnetic layer, Eqs. (34) reduce to usual electrodynamic boundary conditions.

There are, however, two special cases for which directions of the internal and external magnetic fields concide and Eqs. (34) permit analytical solution. The first case is the case of an *in-plane magnetized* film along the anisotropy axis $\widehat{\mathbf{x}}$, for which $\theta_0 = \theta = 0$ and $\phi_0 = \phi = 0$. In this case internal field is equal to

$$H_{||} = H_0 + H_a \ . \tag{35}$$

The second case is the case of *normally magnetized* film, when $\theta_0 = \theta = \pi/2$ (for normal magnetization angles ϕ_0 and ϕ are undetermined) and internal field is given by

$$H_{\perp} = H_0 - 4\pi M_0 \ . \tag{36}$$

This normally-magnetized state is stable only for $H_0 > 4\pi M_0 + H_a$.

3.2. *Equation of motion in complex canonical circularly-polarized variables*

It is convenient to represent magnetization vector in coordinate system, connected with the equilibrium direction $\widehat{\boldsymbol{\zeta}}$. Namely, in addition to the ort $\widehat{\boldsymbol{\zeta}}$, we introduce two unit vectors

$$\widehat{\boldsymbol{\xi}} = -\cos\theta\,\widehat{\mathbf{z}} + \sin\theta\cos\phi\,\widehat{\mathbf{x}} + \sin\theta\sin\phi\,\widehat{\mathbf{y}}\,, \tag{37a}$$

$$\widehat{\boldsymbol{\eta}} = -\sin\phi\,\widehat{\mathbf{x}} + \cos\phi\,\widehat{\mathbf{y}} \ . \tag{37b}$$

Vectors $\widehat{\boldsymbol{\zeta}}$, $\widehat{\boldsymbol{\xi}}$, and $\widehat{\boldsymbol{\eta}}$ form orthonormalized right-handed basis, and magnetization $\boldsymbol{M}$ can be written as

$$\boldsymbol{M} = M_\zeta \widehat{\boldsymbol{\zeta}} + m_\xi \widehat{\boldsymbol{\xi}} + m_\eta \widehat{\boldsymbol{\eta}} \ . \tag{38}$$

Note, that due to conservation of the length of magnetization vector, three components of the magnetization are connected by the relation

$$|\boldsymbol{M}|^2 = M_\zeta^2 + m_\xi^2 + m_\eta^2 = M_0^2 \ . \tag{39}$$

This condition means that there are only two independent components of the magnetization, and, therefore, we can describe magnetization using only *one complex variable* $a = a(\boldsymbol{M})$. The convenient choice of the complex variable a is given by the Holstein–Primakoff transformation

$$a = \frac{m_\xi - i m_\eta}{\sqrt{2M_0(M_0 + M_\zeta)}} \,. \tag{40}$$

In contrast with the standard Holstein–Primakoff transformation, we defined a as dimensionless variable, directly connected with the precession angle $\psi = \arccos(1 - 2|a|^2) \approx 2|a|$. It is clear, that physical meaning have only values of a such that $|a| \le 1$.

The inverse to Eq. (40) transformation is given by

$$\boldsymbol{M} = M_0(1 - 2|a|^2)\widehat{\boldsymbol{\zeta}} + M_0\sqrt{1 - |a|^2}\left[(\widehat{\boldsymbol{\xi}} + i\widehat{\boldsymbol{\eta}})a + (\widehat{\boldsymbol{\xi}} - i\widehat{\boldsymbol{\eta}})a^*\right] \,. \tag{41}$$

One can easily check that Eq. (39) is automatically satisfied with the substitution Eq. (41).

In new complex variable a equation of motion Eq. (27) becomes the perturbed Hamiltonian equation with the Hamiltonian $\mathcal{H} = \gamma W_0/2M_0$:

$$\frac{\partial a}{\partial t} = -i\frac{\delta \mathcal{H}}{\delta a^*} + F_a \,, \tag{42}$$

where the "perturbation force" F_a has the form

$$F_a = \frac{\partial a}{\partial \boldsymbol{M}} \cdot \boldsymbol{T}_\epsilon \,. \tag{43}$$

Thus, the pair of complex-conjugated variables (a, a^*) gives *complex canonical circularly-polarized variables* for the unperturbed Landau–Lifshits equation and one can use all the power of well-developed quasi-Hamiltonian approach for the description of the studied system.

Note that at the equilibrium ($\boldsymbol{M} = M_0\widehat{\boldsymbol{\zeta}}$) $a = 0$, and, respectively, for finite, but not very large precession angles, one can consider a as a small parameter and take into account only first few terms in the series expansion of $\mathcal{H}(a, a^*)$ and $F_a(a, a^*)$. The Laplace operator ∇^2 in the exchange field Eq. (13) can also be considered as a small parameter, since radius R of the nano-contact or nano-pillar is much larger than the exchange length $\lambda_{\rm ex}$ and $\lambda_{\rm ex}\nabla^2 \sim (\lambda_{\rm ex}/R)^2 \ll 1$. Under these simplifications, Hamiltonian $\mathcal{H}$ takes the form

$$\begin{aligned} \mathcal{H} = {} & \mathcal{A}|a|^2 + \frac{1}{2}\left(\mathcal{B}a^2 + \text{c.c.}\right) + \omega_M \lambda_{\rm ex}^2 |\nabla a|^2 \\ & + \left(\mathcal{V}|a|^2 a + \text{c.c.}\right) + \mathcal{U}_1|a|^4 + \left(\mathcal{U}_2|a|^2 a^2 + \text{c.c.}\right) \,, \end{aligned} \tag{44}$$

where c.c. denotes complex-conjugated part and coefficients of the Hamiltonian have the form

$$\mathcal{A} = \omega_H + \frac{1}{2}\cos^2\theta\,\omega_M - \frac{1}{2}\left(\sin^2\theta\cos^2\phi + \sin^2\phi\right)\omega_{\rm an}\,, \tag{45a}$$

$$\mathcal{B} = \frac{1}{2}\cos^2\theta\,\omega_M - \frac{1}{2}\left(\sin\theta\cos\phi - i\sin\phi\right)^2\omega_{\rm an}\,, \tag{45b}$$

$$\mathcal{V} = \cos\theta\left[\sin\theta\,\omega_M + \cos\phi\left(\sin\theta\cos\phi - i\sin\phi\right)\omega_{\rm an}\right]\,, \tag{45c}$$

$$\mathcal{U}_1 = \frac{1}{4}\left[\left(1 - 3\cos 2\theta\right)\left(\omega_M + \omega_{\rm an}/2\right) - 3\cos^2\theta\cos 2\phi\,\omega_{\rm an}\right]\,, \tag{45d}$$

$$\mathcal{U}_2 = -\frac{1}{4}\cos^2\theta\,\omega_M + \frac{1}{4}\left(\sin\theta\cos\phi - i\sin\phi\right)^2\omega_{\rm an}\,, \tag{45e}$$

Here we used the notations $\omega_H = \gamma H$, $\omega_M = 4\pi\gamma M_0$, $\omega_{\rm an} = \gamma H_a$. Note, that in Eq. (44) we dropped constant energy term that does not enter into equations of motion.

3.3. *Diagonalization of the quadratic part of the Hamiltonian*

The quadratic part of the Hamiltonian Eq. (44) (i.e., terms with coefficients $\mathcal{A}$ and $\mathcal{B}$) can be diagonalized by the well-known linear canonical u–v transformation

$$a = ub - vb^*\,, \tag{46}$$

where transformation coefficients u and v are given by

$$u = \sqrt{\frac{\mathcal{A} + \omega_0}{2\omega_0}}\,, \qquad v = \frac{\mathcal{B}^*}{|\mathcal{B}|}\sqrt{\frac{\mathcal{A} - \omega_0}{2\omega_0}}\,, \tag{47}$$

and *ferromagnetic resonance frequency* ω_0 is equal to

$$\begin{aligned}\omega_0^2 &= \mathcal{A}^2 - |\mathcal{B}|^2 \\ &= (\omega_H - \sin^2\phi\,\omega_{\rm an})(\omega_H + \cos^2\theta\,\omega_M) - \sin^2\theta\cos^2\phi\,\omega_H\omega_{\rm an}\,.\end{aligned} \tag{48}$$

Here $\omega_H = \gamma H$, $\omega_M = 4\pi\gamma M_0$, and $\omega_{\rm an} = \gamma H_a$.

The transformation Eq. (46) is a *canonical* transformation, i.e. it conserves the form of Hamiltonian equations, and the equation of motion in terms of b reads

$$\frac{\partial b}{\partial t} = -i\frac{\delta\mathcal{H}}{\delta b^*} + F_b\,, \tag{49}$$

with the same Hamiltonian $\mathcal{H}$ Eq. (44) and the perturbation force F_b

$$F_b = uF_a + vF_a^* \ . \tag{50}$$

The Hamiltonian $\mathcal{H}$, written in terms of b and b^*, has the form

$$\mathcal{H} = \omega_0|b|^2 + D|\nabla b|^2 + \left(\mathcal{W}_1|b|^2 b + \text{c.c.}\right) + \left(\mathcal{W}_2 b^3 + \text{c.c.}\right) + \mathcal{T}|b|^4 \,, \tag{51}$$

where the *dispersion coefficient* D is given by

$$D = (u^2 + |v|^2)\omega_M \lambda_{\text{ex}}^2 = \frac{\mathcal{A}}{\omega_0}\omega_M \lambda_{\text{ex}}^2 \,, \tag{52}$$

and all higher-order coefficients are given by

$$\mathcal{W}_1 = \frac{3}{2}\left(u^2 + |v|^2\right)\left(u\mathcal{V} - v^*\mathcal{V}^*\right) - \frac{1}{2}\left(u\mathcal{V} + v^*\mathcal{V}^*\right) \,, \tag{53a}$$

$$\mathcal{W}_2 = -uv^*\left(u\mathcal{V} - v^*\mathcal{V}^*\right) \,, \tag{53b}$$

$$\mathcal{T} = \frac{1}{2}\left[3\left(u^2 + |v|^2\right)^2 - 1\right]\mathcal{U}_1 - 3u\left(u^2 + |v|^2\right)\left(v\mathcal{U}_2 + v^*\mathcal{U}_2^*\right) \ . \tag{53c}$$

In (51) we neglected non-resonant forth-order terms proportional to b^4 and $|b|^2 b^2$ and non-resonant dispersion terms $\sim (\nabla b)^2$ (and corresponding complex-conjugated terms), which do not lead to any significant effects and can be eliminated by the technique described in the following subsection.

From the physical point of view, complex amplitude b is proportonal to the amplitude of *elliptically polarized* microwave precession.

3.4. *Elimination of non-resonant three-wave processes*

Third-order terms in Eq. (51) describe non-resonant three-wave processes, which can be eliminated by the nonlinear transformation

$$b \approx \sqrt{\frac{\omega_0}{\mathcal{A}}}c + \frac{\omega_0}{\mathcal{A}}\frac{\mathcal{W}_1 c^2 - 2\mathcal{W}_1^*|c|^2 - \mathcal{W}_2^*(c^*)^2}{\omega_0} + O(c^3) \ . \tag{54}$$

The Hamiltonian $\mathcal{H}_c = \mathcal{A}\mathcal{H}/\omega_0$ in c-variables takes very simple form

$$\mathcal{H}_c = \omega_0|c|^2 + D|\nabla c|^2 + \frac{N}{2}|c|^4 \,, \tag{55}$$

where the *nonlinear frequency shift coefficient* N is equal to

$$N = \frac{2\omega_0}{\mathcal{A}}\left(\mathcal{T} - 3\frac{|\mathcal{W}_1|^2 + |\mathcal{W}_2|^2}{\omega_0}\right) \ . \tag{56}$$

In general case expression for N is rather complicated. It should be noted, however, that the nonlinearity coefficient N strongly depends on the direction of the internal magnetic field. Thus, for in-plane magnetized film along the anisotropy axis $\widehat{\mathbf{x}}$ ($\theta = 0$, $\phi = 0$) one obtains

$$N_{||} = -\frac{4\omega_H\omega_M(\omega_H + \omega_M/4) + \omega_{\text{an}}(8\omega_0^2 + 3\omega_M^2)}{4\omega_0(\omega_H + \omega_M/2)}\,, \tag{57a}$$

while for normally magnetized film ($\theta = \pi/2$)

$$N_{\perp} = \frac{8\omega_M(\omega_0^2 + 3\omega_{\text{an}}^2/8) + 4\omega_H\omega_{\text{an}}(\omega_H - \omega_{\text{an}}/4)}{4\omega_0(\omega_H - \omega_{\text{an}}/2)}\,. \tag{57b}$$

One can see that $N_{||} < 0$, while $N_{\perp} > 0$ and, thus, nonlinear frequency shift changes sign when direction of the bias magnetic field changes from the in-plane orientation to the normal to the plane orientation.

The perturbed equation of motion for c can be written as

$$\frac{\partial c}{\partial t} = -i\frac{\delta\mathcal{H}_c}{\delta c^*} + F_{\text{d}} + F_{\text{f}} + F_{\text{s}}\,, \tag{58}$$

where F_{d} is the dissipative force

$$F_{\text{d}} = -\Gamma_0(1 + Q_1|c|^2)c + \kappa\nabla^2 c\,, \tag{59a}$$

F_{f} is the force caused by the perturbation magnetic field $\widetilde{\boldsymbol{h}}$

$$F_{\text{f}} = \lambda_{\text{f}} + V_{\text{f}}c^* - i\Omega_{\text{f}}c\,, \tag{59b}$$

and F_{s} is the current-induced force

$$F_{\text{s}} = \lambda_{\text{s}} + V_{\text{s}}c^* + \Gamma_{\text{s}}\left(1 - |c|^2\right)c\,. \tag{59c}$$

In Eq. (59) we used notations

$$\Gamma_0 = \alpha_{\text{G}}\mathcal{A}\,, \tag{60a}$$

$$Q_1 = 2 - 3\left(\frac{\omega_0}{\mathcal{A}}\right)^2 + \frac{2\mathcal{U}_1}{\mathcal{A}} + \frac{24\text{Re}(\mathcal{B}^*\mathcal{V}^2)}{\omega_0^2\mathcal{A}} + 2\left[3 - \left(\frac{\omega_0}{\mathcal{A}}\right)^2\right]\left[q_1\left(\frac{\omega_0}{\omega_M}\right)^2 - \left(\frac{|\mathcal{V}|}{\omega_0}\right)^2\right], \tag{60b}$$

$$\kappa = \alpha_G \omega_M \lambda_{ex}^2 , \tag{60c}$$

$$\lambda_f = i\sqrt{\frac{\omega_0}{\mathcal{A}}}\Omega_+ , \qquad \lambda_s = -\sqrt{\frac{\omega_0}{\mathcal{A}}}\varepsilon\eta p_+ j , \tag{60d}$$

$$V_f = 2i\left[2\mathcal{B}^*\Omega_{||} + \mathcal{W}_1^*\Omega_+ - \mathcal{W}_2^*\Omega_-\right]/\omega_0 , \tag{60e}$$

$$V_s = -2\varepsilon\eta j\left[\mathcal{W}_1^* p_+ + \mathcal{W}_2^* p_-\right]/\omega_0 ,$$

$$\Omega_f = \left[\mathcal{A}\Omega_{||} + 4\mathrm{Re}(\mathcal{W}_1\Omega_+)\right]/\omega_0 , \tag{60f}$$

$$\Gamma_s = \varepsilon\eta p_{||} j , \tag{60g}$$

where

$$\Omega_+ = \Omega_-^* = \gamma\left(\widetilde{\boldsymbol{h}}\cdot((u-v)\widehat{\boldsymbol{\xi}} - i(u+v)\widehat{\boldsymbol{\eta}})\right)/2 , \; \Omega_{||} = \gamma\left(\widetilde{\boldsymbol{h}}\cdot\widehat{\boldsymbol{\zeta}}\right), \tag{61a}$$

$$p_+ = p_-^* = \left(\widehat{\mathbf{p}}\cdot((u+v)\widehat{\boldsymbol{\xi}} - i(u-v)\widehat{\boldsymbol{\eta}})\right)/2 , \quad p_{||} = \left(\widehat{\mathbf{p}}\cdot\widehat{\boldsymbol{\zeta}}\right) . \tag{61b}$$

In Eq. (59a) Γ_0 is the linear relaxation rate of uniform magnetization precession, $Q_1 \sim 1$ is the dimensionless parameter describing nonlinear relaxation, and term $\kappa\nabla^2 c$ describes increase of the relaxation rate with the increase of the wavevector of the excited spin waves.

Terms λ_f in Eq. (59b) and λ_s in Eq. (59c) describe direct excitation of magnetization precession by magnetic field (λ_f) or by spin-polarized current (λ_s). It is neccessary to account for these terms only when the frequency of the driving force is close to the FMR frequency ω_0, i.e. when external magnetic field and/or current have microwave components. The most important effect of these terms is the phase-locking of current-driven microwave oscillator to a weak external microwave signal.[6,14] In the current paper we will not consider phase-locking effects and these terms will be ignored below.

Terms $V_f c^*$ in Eq. (59b) and $V_c c^*$ in Eq. (59c) describe first-order parametric excitation of spin waves by magnetic field and/or spin-polarized current. These terms are important in the case when magnetic field or current have microwave components at frequency close to $2\omega_0$. In the current paper we will not consider the parametric excitation of spin waves and these terms will be ignored below.

Term $-i\Omega_f c$ in Eq. (59b) describe changes of spin wave resonance frequency under the influence of current-induced Oersted field. Due to the smallness of this frequency change compared to FMR frequency ω_0, we will ignore this effect.

The last term in Eq. (59c) describes *current-induced negative dissipation* induced by the spin-polarized current. This is the main effect caused by the spin-polarized current. As it is clear from Eq. (60g), for certain current polarity ($j > 0$ for collinear orientations of magnetizations of the "free" and "fixed" magnetic layers, $p_{||} > 0$) $\Gamma_s > 0$ and spin-polarized current leads to exponential growth of the spin wave amplitude c. For opposite current polarity $\Gamma_s < 0$ and current increases the spin wave damping rate, in which case spin-polarized current does not lead to any interesting effects. Factor $(1 - |c|^2)$ describes reducing of current efficiency with the increase of the spin wave power $|c|^2$. This factor, together with natural nonlinear dissipation rate $\Gamma_0 Q_1 |c|^2$, limits the growth of the excited spin waves at certain current-dependent level.

4. Magnetization Dynamics in an Isolated Nano-Contact or Nano-Pillar

4.1. *Current-induced magnetization dynamics in macrospin approximation*

In this section we consider current-induced magnetization dynamics in a circular nano-pillar of the radius R under the action of direct bias current. We will assume that both the current distribution and spin wave profile are uniform across the nano-pillar. Such an approximation, that neglects the possibility of spatial variation of the magnetization, is called *macrospin approximation*. In this case spin wave amplitude is function of time only, $c = c(t)$, and its dynamics is governed by the equation

$$\begin{aligned}\frac{dc}{dt} &= -i\omega_0 c - iN|c|^2 c - \Gamma_0(1 + Q_1|c|^2)c + \Gamma_s(1 - |c|^2)c \\ &= -i(\omega_0 + N|c|^2)c - (\Gamma_0 - \Gamma_s)c - (Q_1\Gamma_0 + \Gamma_s)|c|^2 c\,. \qquad (62)\end{aligned}$$

This equation is the equation on nonlinear oscillator with amplitude-dependent frequency $(\omega_0 + N|c|^2)$, linear dissipation $(\Gamma_0 - \Gamma_s)$ (which can either positive or negative, depending on the polarity and magnitude of the bias current), and nonlinear dissipation $(Q_1\Gamma_0 + \Gamma_s)$, which is always positive.

For $\Gamma_s > \Gamma_0$ the total effective linear damping becomes negative, and Eq. (62) has non-trivial stationary solution

$$c(t) = c_0 e^{-i\omega t}\,, \qquad (63)$$

where $\omega = \omega_0 + N|c_0|^2$ and stationary amplitude c_0 is determined from the condition of vanishing total (linear and nonlinear) dissipation:

$$|c_0|^2 = \frac{\Gamma_s - \Gamma_0}{\Gamma_s + Q_1\Gamma_0} \, . \tag{64}$$

Condition of vanishing effective linear damping $\Gamma_s = \Gamma_0$ determines critical current density $j_{th} = \Gamma_0/\varepsilon\eta p_{||}$ or critical current $I_{th} = \pi R^2 j_{th}$, at which microwave generation starts. Introducing *supercriticality* of the current $\zeta = I/I_{th}$ and using Eq. (64) one can write expression for the generated frequency ω in the form

$$\omega = \omega_0 + N\frac{\zeta - 1}{\zeta + Q_1} \, . \tag{65}$$

Equation (65) describes dependence of the generated frequency on bias current and on the magnitude and direction of the bias magnetic field (through the dependence of ω_0 and N on these parameters).

The dependences similar to Eq. (64) and Eq. (65) without account of nonlinear damping ($Q_1 = 0$) were first obtained in Ref. 15 for the case of normally magnetized film and in Ref. 16 for an obliquely magnetization angle.

The generated frequency ω equals FMR frequency $\omega = \omega_0$ at the threshold of generation, $I = I_{th}$, and, with the increase of the current, can either increase (for $N > 0$) or decrease ($N < 0$). The former case ("blue" frequency shift) is observed for magnetization angles, close to the normal to the film direction, while the latter case ("red" frequency shift) corresponds to almost in-plane magnetized "free" magnetic layer.

For moderate currents ($\zeta \leq 1.5$) generated frequency changes approximately linearly with the current with the slope

$$\frac{d\omega}{dI} \approx \frac{N}{(1 + Q_1)I_{th}} \, . \tag{66}$$

4.2. *Current-induced excitation of linear spin wave modes in magnetic nano-contacts*

The considered above macrospin approximation can be used only in the case in magnetic nano-pillars. In the case of current-carrying nano-contact to continuous magnetic layer one can not neglect spatial variations of the microwave magnetization, since in this case variable part of the magnetization should vanish far from the nano-contact.

Here we will consider linear regime of spin wave excitation by dirrect current in a circular nano-contact of the radius R. The linear analysis allows one to determine only the threshold current and frequency, generated at the threshold. We will assume that the current distribution is uniform inside the nano-contact area, $j = I/\pi R^2$ for $r < R$, where I is the total current through the nano-contact, and vanishes outside the nano-contact, $j = 0$ for $r > R$. Neglecting the nonlinear terms, equation of motion for the spin wave amplitude c in the nano-contact area $r < R$ has the form

$$\frac{\partial c}{\partial t} = -i\omega_0 c + iD\nabla^2 c - \Gamma_0 c + \kappa\nabla^2 c + \Gamma_s c \, . \tag{67}$$

Outside the contact the equation of motion has the same form with $\Gamma_s \to 0$.

Partial regular stationary ($\partial/\partial t = -i\omega$) solution of Eq. (67) for $r < R$ in cylindrical coordinates can be written as

$$c(r, \phi) = C_{\text{in}} J_\ell(k_{\text{in}} r) e^{i\ell\phi} \, , \tag{68}$$

where C_{in} is constant of integration, $J_\ell(x)$ is the Bessel function of the first kind, and complex "inner" wavevector k_{in} equals to

$$k_{\text{in}}^2 = \frac{(\omega - \omega_0) + i(\Gamma_0 - \Gamma_s)}{D - i\kappa} \, . \tag{69}$$

Decaying at $r \to \infty$ solution for the outer region $r > R$ can be written as

$$c(r, \phi) = C_{\text{out}} H_\ell^{(1)}(k_{\text{out}} r) e^{i\ell\phi} \, , \tag{70}$$

where C_{out} is another integration constant, $H_\ell^{(1)}(x)$ is the Hankel function of the first kind, and "outer" wavevector k_{out} is given by

$$k_{\text{out}}^2 = \frac{(\omega - \omega_0) + i\Gamma_0}{D - i\kappa} \, . \tag{71}$$

Matching solutions (68) and (70) at the boundary of the nano-contact $r = R$ gives complex transcendenal equation

$$\frac{k_{\text{in}} J_\ell'(k_{\text{in}} R)}{J_\ell(k_{\text{in}} R)} = \frac{k_{\text{out}} H_\ell^{(1)'}(k_{\text{out}} R)}{H_\ell^{(1)}(k_{\text{out}} R)} \, , \tag{72}$$

which implicitly determines two real parameters: threshold current I_{th} and generated at the threshold frequency ω_{th}.

Equation (72) has infinitely many solutions corresponding to excitation of spin wave modes with different azimuthal ℓ and radial n (number of nodes inside the current-carrying area) mode indices. The physical meaning, however, has only the mode with the lowest excitation threshold. This mode corresponds to radially-symmetric $\ell = 0$ nodeless $n = 0$ mode, and only this case will be considered below.

In the conservative limit ($\Gamma_0 = 0$, $\kappa = 0$) lowest solution of Eq. (72) corresponds to $k_{out} = k_0 \simeq 1.2/R$, generated frequency $\omega_{th} = \omega_0 + Dk_0^2$, and threshold current $I_{th} \simeq 5.85D/(\varepsilon\eta p_{||})$. Note that, in contrast with the macrospin approximation, threshold current remains finite even for vanishing dissipation. The physical reason for this is that in the case of nano-contact spin waves, excited in a finite region of magnetic film, propagate out of the active region, which creates effective propagation losses. The energy flux, carried by the propagating spin waves, can be estimated as $|c|^2 \cdot 2Dk_0 \cdot 2\pi R$ — energy density ($|c|^2$) multiplied by spin wave group velocity ($2Dk_0$) and by the perimether of the nano-contact ($2\pi R$). This energy flux should be compensated by the current-induced energy pumping $\Gamma_s|c|^2\pi R^2$, which gives finite value of the threshold current. It is interesting to note that the threshold current in conservative case does not depend on the nano-contact radius R, which is consequence of the parabolic dispersion law for spin waves.

First non-conservative correction of Eq. (72) gives the following expressions for the threshold current and frequency:

$$I_{th} \simeq \frac{\pi R^2}{\varepsilon\eta p_{||}} \left(1.86\frac{D}{R^2} + \Gamma_0 - 0.08\frac{\kappa}{R^2} \right) , \tag{73a}$$

$$\omega_{th} \simeq \omega_0 + 1.43\frac{D}{R^2} + 0.82\Gamma_0 - 1.92\frac{\kappa}{R^2} . \tag{73b}$$

For most experiments on current-induced spin wave generation in small ($R \leq 40$ nm) nano-contacts non-conservative corrections are small compared to zero-order conservative contributions to threshold current and generated frequency.

For the first time the threshold (73) of the linear cylindrical spin wave mode for derived in Ref. 3 for the case of normally magnetized magnetic layer.

4.3. *Current-induced spin wave "bullets" in magnetic nano-contacts*

Analysis of first nonlinear corrections to the linear solution Eq. (73) in the case of normally magnetized film[17] showed that the nonlinear terms do not change qualitatively the nature of the excited spin wave in this case. For supercritical currents $I > I_{th}$ the amplitude of the excited spin wave is saturated at the level $A^2 \propto (I - I_{\rm th})$, and the frequency of the mode is nonlinearly blue-shifted by $\propto NA^2$.

The situation is very different in the case of in-plane magnatized film.[18] In this case nonlinear frequency shift is negative, $N < 0$, and magnetic film supports, in the absence of dissipative and current-induced terms, new types of *nonlinear* excitations — self-localized spin wave "bullets", which have vanishing propagation losses. Interplay between the dissipation and current-induced negative damping can stabilize such bullets leading to a new regime of spin wave excitation.

To show this more rigorously, consider the equation for spin wave amplitude (58) in the nonlinear regime. For simplicity, we will neglect the influence of nonlinear $\Gamma_0 Q_1 |c|^2 c$ and nonuniform $\kappa \nabla^2 c$ contributions to the natural damping. In this case Eq. (58) takes the form

$$\frac{\partial c}{\partial t} = -i(\omega_0 + N|c|^2 - D\nabla^2)c - \Gamma_0 c + \Gamma_{\rm s}\Theta(R_{\rm c} - r)(1 - |c|^2)c\,, \qquad (74)$$

where $\Theta(x)$ is the unit Heaviside function ($\Theta(x) = 1$ for $x > 0$ and $\Theta(x) = 0$ otherwise) and $R_{\rm c}$ is the radius of the nano-contact.

Without damping and current terms ($\Gamma_0 = 0$, $\Gamma_{\rm s} = 0$) Eq. (74) coincides with the well-known (2+1)-dimensional nonlinear Schrödinger equation (NSE).[19] In the considered case of an in-plane magnetized film the nonlinear coefficient N is negative, and the nonlinearity and dispersion satisfy the well-known Lighthill criterion $ND < 0$ (i.e. they act in opposite directions), and the NSE has a nonlinear self-localized radially-symmetric standing solitonic solution (or the solution in the form of a *standing spin wave bullet*)

$$c(t, r) = B_0 \psi(r/\ell) e^{-i\omega t}\,, \qquad (75)$$

where dimensionless function $\psi(x)$, having maximum value of 2.2 at $x = 0$, describes the profile of the bullet. This function is the localized solution of the equation

$$\psi'' + \frac{1}{x}\psi' + \psi^3 - \psi = 0\,, \qquad (76)$$

which has to be found numerically (see e.g. Refs. 20, 21).

In Eq. (75) B_0, ℓ, and ω are the characteristic amplitude, characteristic size, and frequency of the bullet, respectively. Among these three parameters only one is independent. Taking the amplitude B_0 as an independent parameter, we can express the two other parameters as

$$\omega = \omega_0 + NB_0^2\,, \qquad \ell = \frac{\sqrt{|D/N|}}{B_0}\,. \qquad (77)$$

We would like to stress, that the frequency of the spin wave bullet lies *below* the linear frequency ω_0 of the ferromagnetic resonance (see Eq. (77), and note that $N < 0$), i.e. outside the spectrum of linear spin waves. This is the main reason for the self-localization of the spin wave bullet, as the effective wave number of the spin wave mode with frequency (77) is purely imaginary. It also follows from Eq. (75) and the expansion condition $|c| < 1$, that the maximum magnitude of B_0 for which our perturbative approach is still correct is $B_0 = 0.46$.

It is well-known[19] that the bullet-like solutions of (2+1)-dimensional NSE are unstable with respect to the small perturbations: the wave packets having the bullet shape (75), but amplitudes smaller than B_0, decay due to the dispersion spreading, while the wave packets having amplitudes higher than B_0 collapse due to the nonlinearity. At the same time, Eq. (74) with both Gilbert dissipation Γ_0 and current Γ_s is a two-dimensional analog of a Ginzburg–Landau equation that is known to have stable localized solutions (see, e.g., review 22). In our particular case the current-induced terms in Eq. (74) will stabilize spin wave bullets and will prevent them from both the dispersion spreading and nonlinear collapse.

One can assume that for a small damping rate Γ_0 and current I the full non-conservative equation (74) will have a bullet-like solution, only slightly different from the exact solution Eq. (75) of the conservative NSE equation. It is clear, however, that not all of such solutions can be supported in our case. For example, small-amplitude bullets, for which $\ell \gg R_c$, practically do not interact with the spatially-localized current and will decay due to the linear dissipation. The large-amplitude ($B_0 \geq 1$) bullets, on the other hand, will also decay because the effective damping $\Gamma - \Gamma_s(1 - |b|^2)$ for them changes sign and becomes positive.

To find if it is possible to balance the effects of the Gilbert damping and spin-polarized current for a certain amplitude of the excited spin wave mode, we shall consider the time evolution of the mode energy that we shall define as

$$E \equiv \int_0^\infty |c(t,r)|^2 r dr \,, \tag{78}$$

which is conserved in the case of a conservative NSE (when $\Gamma_0 = 0$, $\Gamma_s = 0$). From the full Eq. (74) one can find an exact equation for the time rate of the energy variation dE/dt:

$$\frac{dE}{dt} = -2\Gamma E + 2\Gamma_s \int_0^{R_c} |c|^2 (1 - |c|^2) r dr \,. \tag{79}$$

Assuming that the mode profile is approximately the same as the profile of a bullet, one can substitute Eq. (75) for c in Eq. (79). This gives the following equation

$$\frac{dE}{dt} = 2\left|\frac{D}{N}\right| \left\{\sigma I \left[\eta_2(qB_0) - B_0^2\eta_4(qB_0)\right] - \chi\Gamma\right\} , \tag{80}$$

where

$$\chi \equiv \int_0^\infty \psi^2(x) x dx \approx 1.86 \tag{81}$$

is the constant form-factor of the bullet,

$$q \equiv \sqrt{\left|\frac{N}{D/R_c^2}\right|} \tag{82}$$

is the parameter describing the strength of nonlinearity relative to exchange-originated dispersion for the nano-contact of the radius R_c, and $\eta_n(qB_0)$ is defined as

$$\eta_n(qB_0) \equiv \int_0^\infty f(x/qB_0)\psi^n(x) x dx . \tag{83}$$

The stationary $(dE/dt = 0)$ solution of Eq. (80) exists if

$$\frac{\Gamma_\mathrm{s}}{\Gamma} = \frac{\chi}{\eta_2(qB_0) - B_0^2\eta_4(qB_0)} . \tag{84}$$

This equation implicitly defines the amplitude B_0 (and, therefore, frequency ω and size ℓ) of the stationary bullet as a function of the system parameters (Γ_s/Γ and q).

On the other hand, equation (84) can be interpreted as an equation that defines a current magnitude $I = \pi R_\mathrm{c}^2\Gamma_\mathrm{s}/\varepsilon\eta p_{||}$ which is necessary to support a stationary spin wave bullet of the amplitude B_0. The dependence of the normalized bias current $\zeta(B_0) = I(B_0)/I_\mathrm{min}$ (where $I_\mathrm{min} \equiv \pi R_\mathrm{c}^2\Gamma_0/\varepsilon\eta p_{||}$) for two values of the nonlinearity factor q is shown in Fig. 1. This dependence has a clear minimum corresponding to the amplitude $B_0 = B_\mathrm{th}$ of a bullet formed at the threshold of microwave generation by spin-polarized current.

It is clear from Fig. 1 that in the case of a reasonably large nonlinearity factor $q \geq 3$ the threshold current $I_\mathrm{th} = I(B_\mathrm{th})$, obtained by minimization of Eq. (84), only slightly exceeds the minimum possible value I_min. Above the threshold, for any $I^* > I_\mathrm{th}$, there are two possible stationary amplitudes of the generated spin wave bullet: B^*_min and B^*_max. As usual, the

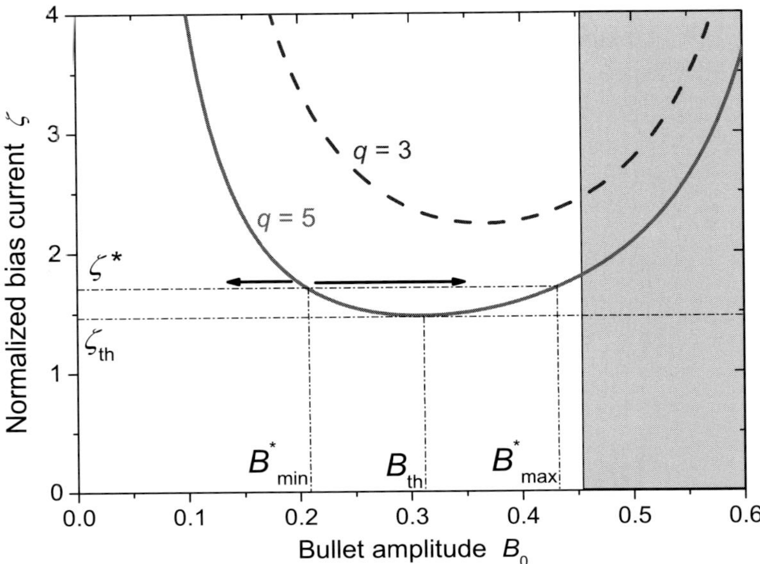

Fig. 1. Dependence (84) of the normalized bias current $\zeta = \varepsilon\eta p_{||}j/\Gamma_0$ on the bullet amplitude B_0 for two values of the nonlinearity factor q: solid line — $q = 5$, dashed line — $q = 3$. Dash-dotted lines indicate the threshold current I_{th}, threshold bullet amplitude B_{th} and low (B^*_{min}) and high (B^*_{max}) bullet amplitudes for a certain supercritical current $I^* > I_{th}$. The shaded area to the right of the vertical line $B_0 = 0.46$ indicates the region where our perturbative approach is not valid.

low-amplitude branch B^*_{min} (for which $B_0 < B_{th}$) is unstable, i.e. for any $B_0 < B^*_{min}$ the mode amplitude will decay to the noise level ($B_0 \to 0$), while for any $B_0 > B^*_{min}$ the mode amplitude will increase to the stable value B^*_{max}, see Fig. 1. Thus, the low amplitude branch in Fig. 1, actually, shows the minimum amplitude, required for the development of the subcritical instability of a finite-amplitude spin wave mode. Our analytical result Eq. (84) is heavily based on the assumption that the profile of the spin wave mode generated at the threshold is approximately the same as the profile of a stationary bullet (75). To check the validity of this assumption we solved Eq. (74) numerically. The results of comparison of the spin wave excitation profiles at the threshold obtained for a typical set of experimental parameters[23] from the analytical solution Eq. (75) (solid line) and numerical solution of Eq. (74) (black dots) are shown in Fig. 2. Both nonlinear solutions have a finite amplitude $|c(0)|^2 \approx 0.36$ at the center of the

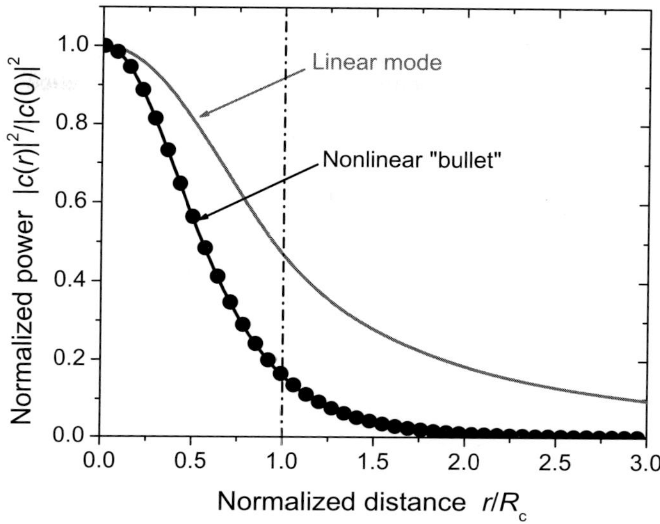

Fig. 2. Normalized profiles of the spin wave mode generated by spin-polarized current at the threshold: solid line — bullet profile (75), circles — result of the numerical solution of Eq. (74), dashed line — profile of the linear eigen-mode calculated from the *linearized* Eq. (74). Vertical dash-dotted line shows the region of current localization. The parameters are: $4\pi M_0 = 16.6$ kG, $H_{\rm app} = H_{\rm ex} = 5$ kOe, $A = 2.85 \cdot 10^{-6}$ erg/cm, $\alpha_{\rm G} = 0.015$, $d = 1.2$ nm, $R_{\rm c} = 25$ nm, $\varepsilon = 0.3$.

nano-contact. One clearly sees that the numerical profile of the nonlinear eigen-mode is practically indistinguishable from the approximate "bullet-like" profile, so the "bullet" model works exceptionally well in this case. For comparison we also present in Fig. 2 the spatial profile of the Slonczhewski-like[3] linear mode, that is obtained from the solution of Eq. (74) where the nonlinear terms (terms containing $|c|^2$) are omitted (dashed line).

In the Fig. 3 we show the dependence of the threshold current on the applied magnetic field. One can see that our bullet model gives quantitative description of the threshold current experimentally measured in Ref. 23, and agrees with experiment much better than the linear threshold Eq. (73). Note, also, that the linear threshold in the in-plane magnetized nano-contact (in contrast with the case of perpendicular magnetization discussed in Ref. 3) demonstrates a non-monotonous dependence on the bias magnetic field H caused by the non-monotonous behavior of the dispersion coefficient D in Eq. (73).

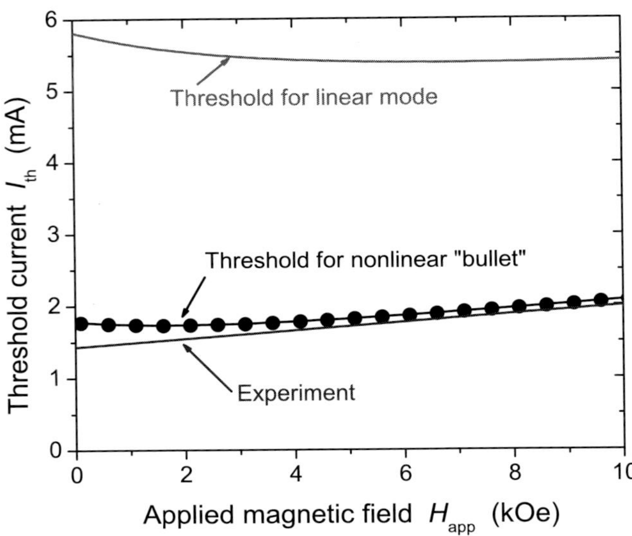

Fig. 3. The dependence of the threshold current I_{th} on the applied magnetic field H_{app}: solid line — nonlinear bullet Eq. (84), dashed line — Slonczewski-like linear mode (73). Dotted line — numerical fit $I(\text{mA}) = 1.43 + 0.056\, H_{\text{app}}(\text{kOe})$ to the experimental data[23] (see Fig. 4 in Ref. 23). The parameters are the same as in Fig. 2.

In the Fig. 4 we demonstrate the comparison of the predictions of our "bullet" model with the results of the experiment[24] (see Fig. 2(a) in Ref. 24) for the magnitude of the spin wave frequency generated at the threshold as a function of the applied magnetic field. It is again clear that the "bullet" model gives a quantitative description of the experiment.

In the Fig. 5 we compare the theoretical dependence of the generated frequency on the bias current calculated from the "bullet" model in the supercritical regime, with the experimental data taken from Ref. 24) (see the inset of Fig. 1(b) in Ref. 24). It can be seen that our theoretical curve is nonlinear and agrees with the experiment, demonstrating linear decrease of frequency with current, only qualitatively. We attribute this to the fact, that our model equation (74) is correct only at a threshold and slightly above it, and a more sophisticated nonlinear model containing higher-order nonlinearities is needed to achieve a full quantitative agreement with experiment in the strongly nonlinear above-threshold regime.

Here we have considered only the *in-plane* magnetized films. Similar results can be obtained for the magnetic films magnetized at small angles

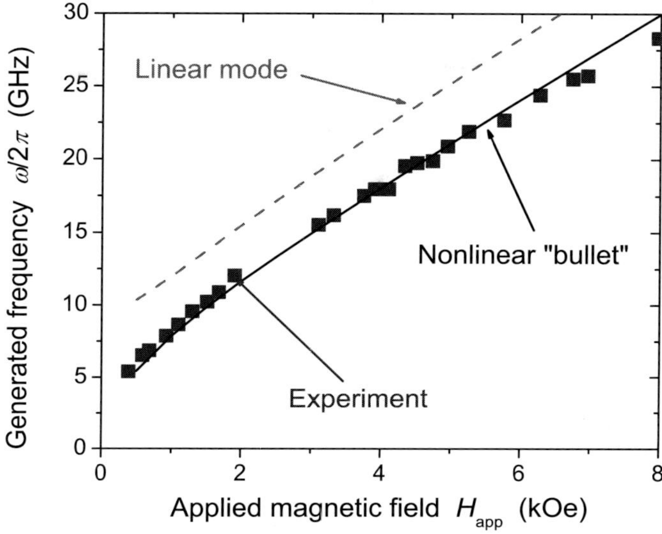

Fig. 4. Dependence of the frequency ω, generated at the threshold, on the applied magnetic field H_{app}. Solid line — frequency of the nonlinear bullet (77), dashed line — frequency of the linear mode (73), symbols — experiment from Fig. 2(a) in Ref. 24. The parameters are: $4\pi M_0 = 8.0$ kG, $H_{\mathrm{app}} = 1$ kOe, $H_{\mathrm{ex}} = 0$, $A = 1.4 \cdot 10^{-6}$ erg/cm, $\alpha_{\mathrm{G}} = 0.02$, $d = 5.0$ nm, $R_{\mathrm{c}} = 20$ nm, $\varepsilon = 0.25$, spectroscopic Lande factor $g = 2$.

to their surface, for which the nonlinear frequency shift is still negative and Lighthill criterion $ND < 0$ is satisfied (see Fig. 8 in Ref. 16 for details).

When the free layer of a magnetic nano-contact is magnetized at an arbitrary angle to the plane of the layer, then, for sufficiently large magnetization angles (angles between the direction of the external bias magnetic field and the film plane) the coefficient of the nonlinear frequency shift changes its sign from negative to positive (see Eq. (56) and the conditions of existence of a self-localized nonlinear spin wave "bullet", having the frequency lower than the FMR frequency of the layer (see Eq. (77)) are violated. For such large magnetization angles only quasi-linear propagating spin wave modes with the frequency that is higher than the FMR frequency in the layer (see Refs. 3 and 17 for details) could be excited in an isolated magnetic nano-contact by the spin-polarized current. Therefore, with the increase of the magnetization angle the character of the spin wave mode excited by the spin -polarized current can change abruptly and the corresponding jumps in the generated frequency can be experimentally

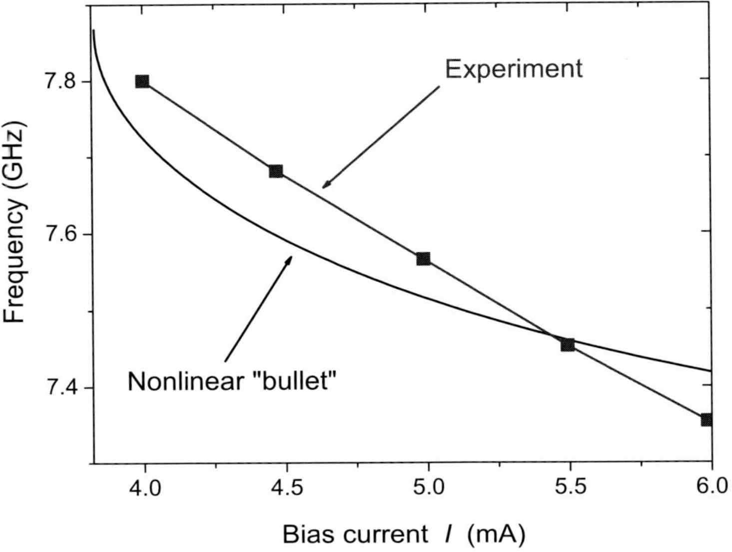

Fig. 5. The dependence of the generated frequency on the bias current for $H_{\text{app}} =$ 1 kOe: solid line — nonlinear bullet, symbols — experiment from the inset of Fig. 1 in Ref. 24. The parameters are the same as in Fig. 4.

detected. The detailed calculations corresponding to this case of oblique magnetization are presented in our paper.[25]

5. Conclusions

In conclusion, we developed a simple analytic theory of microwave excitations in magnetic nano-sized systems caused by the spin-torque effect. The theory is developed using a perturbative approach, and is quantitatively correct only for sufficiently small precession angles (smaller than 60 deg). Nevertheless, this simple theory based on the equation of a nonlinear auto-oscillator (62) gives a qualitatively correct description of the phenomenon and adequately describes most of the currently available experimental data. We hope that the approximate analytic theory developed in this Chapter will provide a good foundation for the development of a practically useful engineering theory of active microwave devices based on the spin-torque effect.

Acknowledgments

This work was supported in part by the MURI Grant No. W911NF-04-1-0247 from the U.S. Army Research Office, by Contract No. W56HZV-07-P-L612 from the U.S. Army TARDEC, RDECOM, by Grant No. ECCS-0653901 from the National Science Foundation of the USA, and by the Oakland University Foundation.

References

1. J. C. Slonczewski, Current-driven excitation of magnetic multilayers, *J. Magn. Magn. Mat.* **159**, L1-L7 (1996).
2. L. Berger, Emission of spin waves by a magnetic multilayer traversed by a current, *Phys. Rev. B* **54**, 9353-9358 (1996).
3. J. C. Slonczewski, Excitation of spin waves by an electric current, *J. Magn. Magn. Mat.* **195**, L261-L268 (1999).
4. V. S. L'vov, *Wave Turbulence Under Parametric Excitation* (Springer-Verlag, New York, 1994).
5. A. G. Gurevich and G.A. Melkov, *Magnetization Oscillations and Waves* (CRC Press, New York, 1996).
6. A. N. Slavin and V. S. Tiberkevich, Nonlinear self-phase-locking effect in an array of current-driven magnetic nanocontacts, *Phys. Rev. B* **72**, 092407 (2005).
7. T. L. Gilbert, A Lagrangian formulation of the gyromagnetic equation of the magnetization field, *Phys. Rev.* **100**, 1243 (1955).
8. V. Tiberkevich and A. Slavin, Nonlinear phenomenological model of magnetic dissipation for large precession angles: Generalization of the Gilbert model, *Phys. Rev. B* **75**, 014440 (2007).
9. K. Yu. Guslienko, S. O. Demokritov, B. Hillebrands, and A. N. Slavin, Effective dipolar boundary conditions for dynamic magnetization in thin magnetic stripes, *Phys. Rev. B* **66**, 132402 (2002).
10. K. Yu. Guslienko and A. N. Slavin, Boundary conditions for magnetization in magnetic nanoelements, *Phys. Rev. B* **72**, 014463 (2005).
11. S. I. Kiselev, J. C. Sankey, I. N. Krivorotov1, N. C. Emley, R. J. Schoelkopf, R. A. Buhrman, and D. C. Ralph, Microwave oscillations of a nanomagnet driven by a spin-polarized current, *Nature* **425** 380 (2003).
12. K. Y. Guslienko, R. W. Chantrell, and A. N. Slavin, Dipolar localization of quantized spin-wave modes in thin rectangular magnetic elements, *Phys. Rev. B* **68**, 024422 (2003).
13. C. Bayer, J. Jorzick, S. O. Demokritov, B. Hillebrands, R. Kouba, R. Bozinoski, A. N. Slavin, K. Guslienko, D. Berkov, N. Gorn, and M. P. Kostylev, Spin-wave excitations in finite rectangular elements of $Ni_{80}Fe_{20}$, *Phys. Rev. B* **72**, 064427 (2005).

14. W. H. Rippard, M. R. Pufall, S. Kaka, T. J. Silva, S. E. Russek, and J. A. Katine, Injection locking and phase control of spin transfer nano-oscillators, Phys. Rev. Lett. **95**, 067203 (2005).
15. S. M. Rezende, F. M. de Aguiar, and A. Azevedo, Spin-wave theory for the dynamics induced by direct currents in magnetic multilayers, *Phys. Rev. Lett.* **94**, 037202 (2005).
16. A. N. Slavin and P. Kabos, Approximate theory of microwave generation in a current-driven magnetic nanocontact magnetized in an arbitrary direction, *IEEE Trans. Magn.* **41**, 1264-1273 (2005).
17. M. A. Hoefer, M. J. Ablowitz, B. Ilan, M. R. Pufall, and T. J. Silva, Theory of magnetodynamics induced by spin torque in perpendicularly magnetized thin films, *Phys. Rev. Lett.* **95**, 267206 (2005).
18. A. Slavin and V. Tiberkevich, Spin wave mode excited by spin-polarized current in a magnetic nanocontact is a standing self-localized wave bullet, *Phys. Rev. Lett.* **95**, 237201 (2005).
19. N. N. Akhmediev and A. Ankiewicz, *Solitons. Nonlinear Pulses and Beams* (Chapman & Hall, London, 1997).
20. R. Y. Chiao, E. Garmire, and C. H. Townes, Self-trapping of optical beams, *Phys. Rev. Lett.* **13**, 479 (1964).
21. Y. Silberberg, Collapse of optical pulses, *Opt. Lett.* **15**, 1282 (1990).
22. I. S. Aranson and L. Kramer, The world of the complex Ginzburg–Landau equation, *Rev. Mod. Phys.* **74**, 99 (2002).
23. W. H. Rippard, M. R. Pufall, and T. J. Silva, Quantitative studies of spin-momentum-transfer-induced excitations in Co/Cu multilayer films using point-contact spectroscopy, *Appl. Phys. Lett.* **82**, 1260 (2003).
24. W. H. Rippard, M. R. Pufall, S. Kaka, S. E. Russek, and T. J. Silva, Direct-current induced dynamics in $Co_{90}Fe_{10}/Ni_{80}Fe_{20}$ Point Contacts, *Phys. Rev. Lett.* **92**, 027201 (2004).
25. G. Gerhart, E. Bankowski, G. A. Melkov, V. S. Tiberkevich, and A. N. Slavin, Angular dependence of the microwave-generation threshold in a nanoscale spin-torque oscillator, *Phys. Rev. B* **76**, 024437 (2007).

INDEX